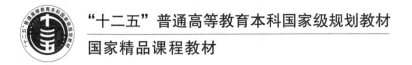

"十二五"普通高等教育本科国家级规划教材

国家精品课程教材

数字图像处理实习教程

（第三版）

贾永红　张　谦　崔卫红　余　卉　编著

武汉大学出版社

图书在版编目(CIP)数据

数字图像处理实习教程/贾永红等编著.—3版.—武汉:武汉大学出版社,2016.11
"十二五"普通高等教育本科国家级规划教材　国家精品课程教材
ISBN 978-7-307-17594-5

Ⅰ.数… Ⅱ.贾… Ⅲ.数字图像处理—实习—高等学校—教材
Ⅳ.TN911.73-45

中国版本图书馆 CIP 数据核字(2016)第 030817 号

责任编辑:王金龙　　　责任校对:汪欣怡　　　版式设计:马　佳

出版发行:武汉大学出版社　　(430072　武昌　珞珈山)
　　　　　(电子邮件:cbs22@whu.edu.cn　网址:www.wdp.com.cn)
印刷:湖北恒泰印务有限公司
开本:787×1092　1/16　印张:21.5　字数:508 千字
版次:2007 年 1 月第 1 版　　2012 年 1 月第 2 版
　　2016 年 11 月第 3 版　　2016 年 11 月第 3 版第 1 次印刷
ISBN 978-7-307-17594-5　　定价:39.00 元

版权所有,不得翻印;凡购买我社的图书,如有质量问题,请与当地图书销售部门联系调换。

前　言

　　自1980年原武汉测绘科技大学航测系开设该课程以来，随着计算机技术的发展和教学条件的改善，经过30多年的建设与改革，已建立起凸显课程基础理论、技术方法和应用三大特征的理论与实践并重的新教学体系。课程性质从当初的任选课转变为限选课，从限选课发展为专业基础必修课，目前已跨入主干课系列；教学模式从最初纯理论发展为"理论教学+课间实习"模式，并改革成现在的"理论教学+课间实习+集中实习"模式；教材建设经历了最初自翻译材料→引用正式出版的翻译教材→拥有自主版权的自编公开出版教材→校"十五"规划教材《数字图像处理》→国家"十一五"→"十二五"规划教材，成为国内50多所知名高校选用的教材，曾荣获全国测绘优秀教材二等奖和高校畅销书一等奖；教学手段从传统教学转变为全程教学现代化。该课程由2003年武汉大学、湖北省精品课程建设成2004年国家精品课程、2013年国家级精品资源共享课。这一切成果的取得，凝聚着课程组负责人和成员的辛劳和汗水，更离不开学院、学校、湖北省教育厅和国家教育部给予的鼓励和支持。

　　随着计算机和信息技术的发展，"数字图像处理实习"现已成为"数字图像处理"课程不可缺少的重要教学环节。为达到理论教学与实践教学并重，我们一直致力于实习教学体系和教学内容建设与改革。建立了"数字图像处理实习"课程独立设课的新教学体系；实习内容涵盖专业基础和专业知识，由浅入深，由单项到综合逐步提升；取材上力求做到先进、新颖、实用，理论联系实际；技能培养由点及面到系统，摆脱简单的知识学习、理论验证等传统实践教学模式，积极引导学生独立钻研和相互协作，努力培养学生设计开发、实践创新和综合运用知识的能力。2007年出版武汉大学"十一五"规划教材《数字图像处理实习教程》，作为国家精品课程实践篇教材，在武汉大学遥感学院本科数字图像处理实习教学环节连续使用了5届，不仅深受学生的欢迎，而且深受图像处理爱好者的喜爱。该书确实对图像处理初学者的动手能力培养起到了重要的启蒙和指导作用；但对有一定数字图像基础的本科生、研究生和科研工作者来说，现有教材还存在欠实用性和通用性，亟待补充和修改。为此，2012年修订版增加了图像内存映射、大幅面图像分块处理技术、GDAL、OpenCV、OpenMP、OpenRS及其应用等；增加、修改和补充实习算法，使算法更加规范、通用和实用。经过近三年应用，受到多方面好评，2014年入选"十二五"普通高等教育本科国家级规划教材。

　　为了更好地满足教学和技术开发的需求，在国家精品课程、国家精品共享课和国家"十二五"规划教材等项目经费资助下，我们着手对《数字图像处理实习教程》（第二版）进行改编，对全书内容进行了补充和完善，作为国家"十二五"规划教材《数字图像处理实习教程》（第三版），奉献给读者。

前言

  本书由四部分内容组成。第一部分是基础篇，包括数字图像处理实习平台搭建、位图文件存取与显示功能、图像内存映射和2个单元课间实习。第二部分是提高篇，包括GDAL、OpenCV、OpenMP、OpenRS及其应用，以及大幅面图像分块处理技术等。通过学习和应用这些资源和技术，提高学生高效开发、设计图像处理算法的能力，达到算法实用、通用。第三部分是综合设计，由3个单元的基础实习和20个具有综合性、应用性和创新性的实习单元组成。集中实习的目的是要求学生进一步掌握和巩固所学图像处理技术的基本原理与方法，理论联系实际，灵活应用所学知识解决实际问题，提高分析问题和解决问题的能力，培养学生的创新能力。第四部分给出了实习内容的相关源代码，仅供参考。另外，每个实习单元后面都附有思考题。

  全书由贾永红策划、设计和组织编写，张谦、姬翠翠等参与第一、二部分有关内容的撰写；崔卫红主要参与第一、三部分有关内容的撰写；余卉、谢志伟、吕臻、祝梦花和周明婷主要参与第三、四部分有关内容的撰写与程序整理；还有李芳芳、潘鹏、马云霞、胡静、高振宇、张岱、余伟等人参与编写实习单元程序代码和流程图绘制。其他内容及全书的统稿、定稿由贾永红完成。

  本书不仅可以作为遥感科学与技术、计算机科学与技术、光学、电子、测绘工程、地理信息系统、通信和自动控制等专业的学生实习教材和参考资料，也可作为工程技术人员和科研人员进行数字图像处理研究和开发的技术资料。

  本书被遴选为国家"十二五"规划教材，完全是集体努力的结晶。初稿曾承蒙关泽群教授和张治国博士审阅斧正。该书引用了一些作者的论文和资料，在此一并表示衷心感谢。

  由于本人水平所限，书中不足之处，恳请读者批评指正。

编 者

2016年1月

目 录

第一部分 基 础 篇

第一章 实习预备知识 ··· 3
第一节 创建 VC++应用工程的基本流程 ····························· 3
第二节 应用程序中添加菜单、对话框资源及其消息处理函数的方法 ········ 18

第二章 位图操作 CDib 类的实现 ··· 33
第一节 位图文件格式与调色板原理 ································· 33
第二节 位图操作类 CDib 的实现 ··································· 34

第三章 基于 CDib 类的位图文件读取、显示和存储 ······················· 51
第一节 位 图 读 取 ··· 51
第二节 位图的显示 ··· 53
第三节 位图的存储 ··· 54

第四章 内存映射技术在大幅面图像读写的应用 ··························· 56
第一节 内存映射文件技术 ··· 56
第二节 基于内存映射文件的文件读取示例 ··························· 58

第五章 课间实习 ··· 60
实习一 灰度图像直方图统计 ······································· 60
实习二 图像增强操作 ··· 61

第二部分 提 高 篇

第六章 GDAL 开源库及其应用 ··· 67
第一节 GDAL 开源库简介 ·· 67
第二节 GDAL 开源库安装及其应用 ·································· 73

第七章 OpenCV 开放源代码简介及其应用 ································· 88
第一节 OpenCV 开放源代码简介 ···································· 88
第二节 OpenCV 库的安装、配置与应用实例 ·························· 89

第八章 大幅面图像分块处理 ……………………………………… 94
第一节 分块处理方法 ……………………………………… 94
第二节 分块处理算法实例 ………………………………… 95

第九章 图像并行处理技术 ……………………………………… 101
第一节 OpenMP 简介与应用实例 ………………………… 101
第二节 基于 MPI 与 OpenMP 混合并行计算 …………… 109
第三节 GPU 并行计算简介及应用 ……………………… 110

第十章 OpenRS 简介与应用 …………………………………… 118
第一节 OpenRS 简介与安装 ……………………………… 118
第二节 OpenRS 应用示例 ………………………………… 123

第三部分 综合设计

第十一章 基础实习(必做) ……………………………………… 131
实习一 实现 RAW→BMP 格式的转换 ………………… 131
实习二 灰度图像对比度增强 …………………………… 134
实习三 图像局部处理 …………………………………… 135

第十二章 综合性实习(选做) …………………………………… 138
实习一 灰度图像中值滤波 ……………………………… 138
实习二 图像几何处理:图像平移、缩放和旋转变换 … 139
实习三 图像频域处理 …………………………………… 145
实习四 伪彩色增强 ……………………………………… 150
实习五 基于高通滤波的影像融合 ……………………… 153
实习六 基于 HIS 变换的影像融合方法 ………………… 155
实习七 灰度图像边缘检测 ……………………………… 157
实习八 图像二值化:判断分析法 ……………………… 158
实习九 Hough 变换检测直线 …………………………… 161
实习十 霍夫曼编码 ……………………………………… 163
实习十一 图像的行程编码 ……………………………… 165
实习十二 纹理图像的自相关函数分析法 ……………… 168
实习十三 灰度共生矩阵特征提取 ……………………… 169
实习十四 基于灰度的模板匹配 ………………………… 172
实习十五 基于特征的模板匹配 ………………………… 174
实习十六 形状特征提取 ………………………………… 175

实习十七　色彩平衡 …………………………………………………………… 176
实习十八　点特征提取 …………………………………………………………… 178
实习十九　图像 K 均值聚类 …………………………………………………… 180
实习二十　分水岭分割 …………………………………………………………… 182

第四部分　实习项目的 C++ 源程序代码

参考文献 ………………………………………………………………………………… 336

第一部分 基 础 篇

数字图像处理实习的目的是要求学生在掌握数字图像处理基本原理与方法的基础上,理论联系实际,灵活应用所学知识解决实际问题,提高分析问题和解决问题的能力,培养学生的创新能力。为此,学生应该对 Visual C++开发工具有所了解,特别是要掌握如何利用 VS 平台搭建起一个 Visual C++应用工程,以及如何实现图像处理常见的基本操作。下面介绍创建 Visual C++应用工程的基本流程,希望学生完成本部分的阅读和操作后,能创建一个基于 MFC 的 Visual C++数字图像处理工程,并能实现位图读写、显示的操作,为顺利完成后面的各项实习和将来能独立地开发应用系统打好基础。

第一章 实习预备知识

为了在 Windows 系统下采用 Visual C++设计图像处理算法,实习前,我们要求学生能用 Visual C++创建应用工程,以及掌握实现图像处理常见的基本操作。

第一节 创建 VC++应用工程的基本流程

对 Visual C++的初学者,按照下列步骤操作,就可初步掌握以 VS 为平台设计 Visual C++应用程序的基本操作流程。

一、创建基于 Visual Studio 6.0 的 Visual C++应用程序的步骤

(1) 打开 Visual C++开发工具软件。
(2) 点击开始/程序/Microsoft Visual Studio 6.0/Microsoft Visual C++ 6.0(如图 1-1 所示)。

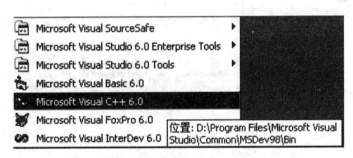

图 1-1 打开 Visual C++开发工具

(3) 创建一个新的项目。

在 Visual C++开发环境下,选择"File"菜单下的"New"菜单,点击进入"New"对话框,选择"Projects"中的"MFC AppWizard(exe)",如图 1-2 所示,在路径"E:\教学\"下创建工程"ImageProcessEx"。也可点击右侧按钮 ... 选择工作路径,根据自己的学号或需要输入工程名称。

(4) 单击"OK"按钮,进入"MFC AppWizard"(MFC 应用程序向导)对话框,如图 1-3 所示,根据提示依次进行选择(共 6 步),直到最后出现"Finish"。

在"MFC AppWizard-Step 1"对话框中,三个单选项分别表示所构建的程序是基于单文档(如 windows 的 notebook)、多文档(如 Microsoft Word)还是基于对话框形式。复选框是文

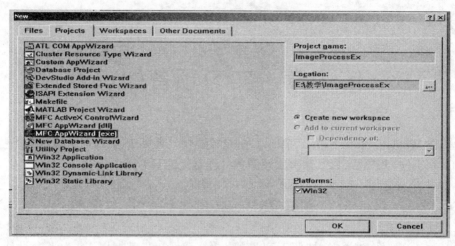

图 1-2 "New"对话框

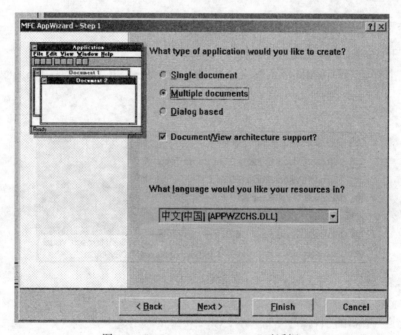

图 1-3 "MFC AppWizard-Step1"对话框

档/视图结构支持(对话框结构时不支持)选项。下面的资源语言选择项,用来为你的程序选择不同的资源语言。若选择了英语[美国][APPWNENU.DLL]等其他资源语言,则在后面添加的菜单、对话框等资源中将不能正常显示中文。

图 1-4 所示"MFC AppWizard-Step 2"对话框用来为你的工程选择是否需要数据库支持以及选择什么样的数据库和数据源。这里不做详细介绍,如感兴趣请查阅有关 Visual C++ 数据库编程的资料。

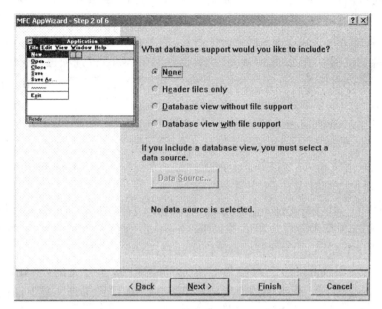

图 1-4 "MFC AppWizard-Step2"对话框

图 1-5 所示的"MFC AppWizard-Step 3"对话框包含"compound document support(复合文档支持)"选项和其他高级支持选项,可针对用户的具体要求将应用程序做成特定的类型。

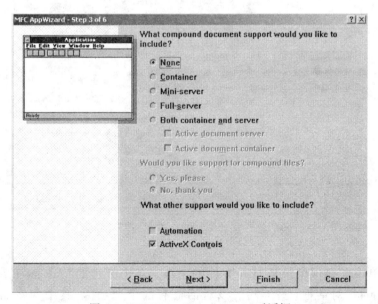

图 1-5 "MFC AppWizard-Step 3"对话框

将文本和图表同时保存在一个文档中,这样的文档就是一种复合文档。只由文本组成的文档不能称为复合文档。如果应用程序不需要设计成复合文档支持型,就选择"None"选

5

项;如果希望将应用程序做成一个全服务器的话,可以选择"Full-server"选项;如果希望将应用程序做成一个容器的话,可以选择"Container"选项。所谓"容器"就是可以嵌入其他对象的应用程序。比如 Microsoft Word 就是一个容器,因为在 Word 中以嵌入的方式可加入一个位图对象,能容纳位图对象,因此称 Word 为容器。"Mini-server"选项也是一个服务器,满足其他应用,但它不能自己独立运行,必须由其他应用来启动运行。而与之相对的"Full-server(完全服务器)"是可以单独运行的。

位于"MFC AppWizard-Step3"对话框下方的其他支持选项,一个是"Automation(自动)",与组件对象模型(Component Object Model,COM)关系密切。如果你要创建一个 COM 组件,又想在网络上发布,那么最好选上该项。另一个是"ActiveX Controls(ActiveX 控件)",是指创建的工程支持使用 ActiveX 组件。

图 1-6 给出的"MFC AppWizard-Step 4"对话框用来设计应用工程的一些外观特征,即用户界面功能。其选项都十分直观明了,如"Docking tollbar(停泊工具栏)"、"Initial status bar(初始状态栏)"、"Print and print preview(打印和打印预览)"、"Context-sensitive Help(上下文相关帮助)"、"3D controls(三维控件)"以及与网络编程有关的"MAPI"和"Windows sockets(Windows 套接)"。再下面的选项是 toolbar 的外观样式选择。最下面的选项是设定

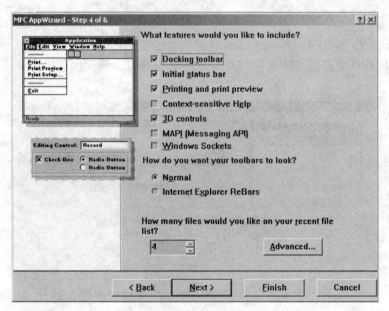

图 1-6 "MFC AppWizard-Step 4"对话框

文件打开菜单中需要保留最近打开文件的个数。点击旁边的"Advanced..."按钮,弹出"Advanced Option"对话框(图 1-7)。该对话框有两个选项卡,其中"Document Template String(文档模板字符串)"选项是对应用程序文档的一些说明。这里有 7 个编辑框,都可以进行修改,不过与其对它们进行修改,不如在生成应用程序之前给定应用程序的名称,这些编辑框中的值都会以字符串的形式各自保存在一个字符资源之中。其中第一个编辑框中的

字符串代表应用程序的扩展名,如希望应用程序的扩展名为".bmp",就可将第一个文本框"File extension(文件扩展名)"中的内容写为"bmp"。"Advanced Option"对话框的第二个选项"Window Styles(窗口类型)"是为选择合适的应用程序外观而设计的,如果希望应用程序一启动就具有最大化的特征,占据整个的桌面空间,那么就选择"Maximized(最大化)"复选框。其他复选框有关于拆分窗口等多个内容,试着修改这些属性,注意应用程序的外观所发生的变化。

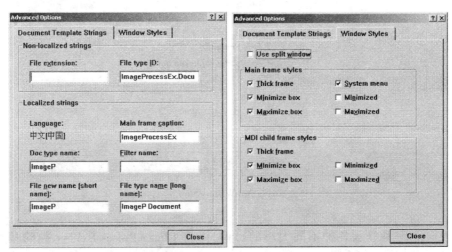

(a)"Document Template Strings"选项卡　　(b)"Window Styles"选项卡

图 1-7　"MFC AppWizard-Step 4"中的"Advanced Options"对话框

　　图 1-8 是"MFC AppWizard-Step 5"对话框,包含应用程序选择界面样式、是否需要为应用程序产生源文件注释和应用程序将如何使用 MFC 库三项内容。其中,应用程序界面样式可以有"MFC Standard(MFC 标准)"和"Windows Explorer(Windows 资源管理器)"两个可选项。在 MFC 库使用的选项中,可以选择"As a shared DLL(作为一个共享的动态链接库)"或"As a statically linked library(作为静态链接库)"。动态链接库与静态链接库的区别在于:动态链接在应用程序运行时才去真正调用动态链接库中的函数;在编译过程中,动态链接库并不直接挂接在应用程序中,这样既节约计算机资源又减少了应用程序的代码量;它可以为多个应用程序使用,但当一个应用程序要使用某个动态链接库中的函数,则必须保证该动态链接库存在,否则程序无法运行。同样,当发布应用程序时,也要将使用的动态链接库一同发布,否则用户无法使用应用程序。静态链接是在使用到某类及其成员函数时,要对类有明确定义,并且在编译过程中要将它们与应用程序紧密地链接在一起。所以,静态链接存在的最大问题是无论在哪个应用程序中使用,都必须有类的源代码,否则便无法完成静态链接。这不仅使程序变得冗长,同时也浪费计算机资源。因此,选择静态链接库方式生成的应用程序将比使用动态链接库方式的应用程序大许多。

　　MFC AppWizard 的最后一步是对应用程序框架所定制生成的类的一个总结。根据前面几个步骤中的选择,呈现工程所得到类的情况,还可以修改工程中存在的类的基类。比如,

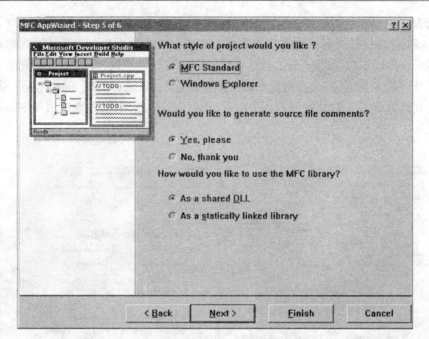

图 1-8 "MFC AppWizard-Step 5"对话框

为了使视窗能完整显示各种不同大小的影像,可将 CImageProcessExView 类的基类设为 CScrollView(滚动视窗类)。如图 1-9 所示。

单击"Finish"按钮,可以看到集成开发环境列出了工程的类型、工程中包含的类以及所选界面外观特征等信息,如图 1-10 所示。点击"New Project Information"对话框的"OK"按钮后,开发环境便根据 MFC AppWizard 步骤 1~6 的选择创建生成工程源程序,其工程开发环境界面如图 1-11 所示。

(5) 自动生成的几个重要类及其功能说明。

CImageProcessExApp 为应用类,负责启动主线程,一般不在这个类工作。一个应用类有且只能有一个从该类派生出来的全局对象,MFC 默认状态下将它命名为 theApp。

CImageProcessExDoc 为文档类,负责对文档内容的管理,包括文件读入与保存等许多与存储相关的操作。永久保存数据操作、文档更新操作以及文档读取操作等都是通过该类完成的,用它来管理数据非常方便,因此在编写应用程序时,常常将该类作为数据的存取基地。

CImageProcessExView 为视图类,负责数据显示,该类的许多方法与图形操作有关。在文档类中保存的数据一般是由视图类来显示给用户的,用户通过视图与文档类中保存的数据进行交互操作。

CMainFrame 为框架类,负责管理整个框架,在多文档应用程序中具有多个视图,而且允许一次显示多个视图。若要求从一个视图切换到另一个视图,同时还允许修改客户区大小和显示多个视图,这使得管理框架为客户区提供的区域变得复杂,可利用 CMainFrame 实现。

CChildFrame 为框架类,负责对视图的管理。

(6) 点击图标 ▨ 对程序进行编译连接,编译无误,再点击 ❗ 运行程序。

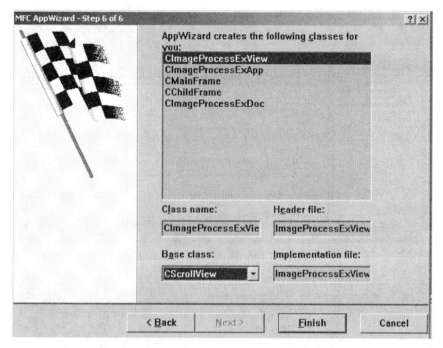

图 1-9 "MFC AppWizard-Step 6"对话框

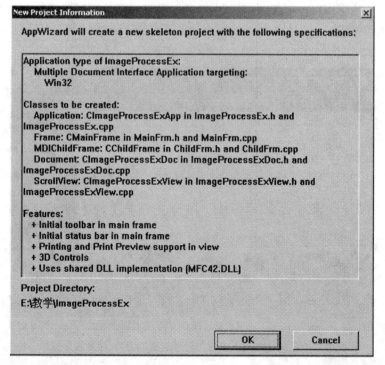

图 1-10 "New Project Information"对话框

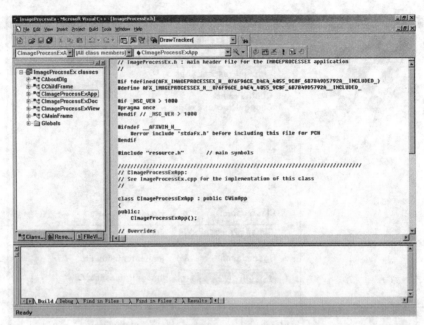

图 1-11 工程开发环境界面

按照以上步骤操作,运行程序,其结果如图 1-12 所示。这是在默认情况下由 MFC 应用向导所创建的空的应用程序框架,其外观与 Word 相似。若要实现图像处理的功能,则需要添加相应资源和代码。

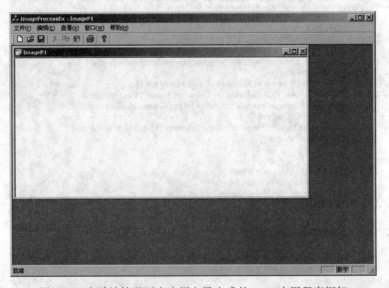

图 1-12 在默认情况下由应用向导生成的 MFC 应用程序框架

二、以 VS2010 为平台创建应用工程的基本流程

Visual Studio 是目前最流行的 Windows 平台应用程序的集成开发环境,从最早的 Visual Studio97,到 2014 年 11 月微软最新发布 Visual Studio 2015,共有 10 个版本。VS2010 及以后的版本都不错。VS2012 版本提供了更为简便优化的界面,实现轻松导航项目应用程序;加入了针对 Windows 8 项目的可视化的工具集,对于 Web 开发,增加了最新的模板、工具以及对 HTML5 和 CSS3 等新标准的全面支持等。荣获在软件界有奥斯卡奖之称的 Jolt 奖 2013 年生产力奖。VS2013 新增了代码信息指示、团队工作室、身份识别、.NET 内存转储分析仪、敏捷开发项目模板、Git 支持以及更强有力的单元测试支持。Visual Studio 2015 预览包含许多新的和令人兴奋的功能,以支持跨平台移动开发、web 和云开发、IDE 生产力增强。鉴于学院实验室目前安装的是 VS2010,下面就给出以 VS2010 为平台创建应用工程的基本流程,其他版本创建应用工程的基本流程与其相似。

(1)打开 Visual Studio 2010 开发工具软件。
(2)点击"开始"→程序→"Microsoft Visual Studio 2010"。
(3)创建一个新的项目。

菜单栏 File→New→Project,弹出 New Project 对话框,我们可以选择工程类型。如果安装完 VS2010 以后第一启动时已经设置为 VC++,则 Installed Templates→Visual C++项会默认展开,而如果没有设置 VC++,则可以展开到 Installed Templates→Other Languages→Visual C++项。因为我们要生成的是 MFC 程序,所以在"Visual C++"下选择"MFC",对话框中间区域会出现三个选项:MFC ActiveX Control、MFC Application 和 MFC DLL。MFC ActiveX Control 用来生成 MFC ActiveX 控件程序。MFC Application 用来生成 MFC 应用程序。MFC DLL 用来生成 MFC 动态链接库程序。我们要选择 MFC Application。在对话框下部有 Name、Location 和 Solution name 三个设置项。意义如下:Name—工程名,Location—解决方案路径,Solution name—解决方案名称。如图 1-13 所示,Name 设为"ImageProcess",Location

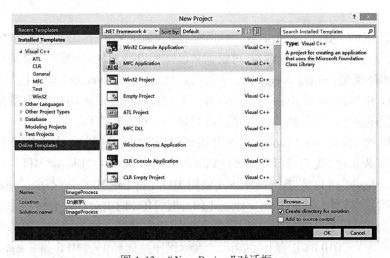

图 1-13 "New Project"对话框

设置为"D:\教学\"，Solution name 默认和 Name 一样，当然可以修改为其他名字，单击"OK"按钮。

（4）单击"OK"按钮，进入"MFC Application Wizard"对话框，如图 1-14 所示，根据提示依次进行选择，直到最后出现 Finish。

图 1-14 所示的"Welcome to the MFC Application Wizard"对话框，下面显示了当前工程的默认设置。如果此设置内容满足工程设计要求，则可以直接点击"Finish"创建工程，也可以点击"Next"对工程进行逐步设置。第一条"Tabbed multiple document interface（MDI）"指此工程是多文档应用程序。如果这时直接单击下面的"Finish"按钮，则可生成具有上面列出设置的多文档程序。

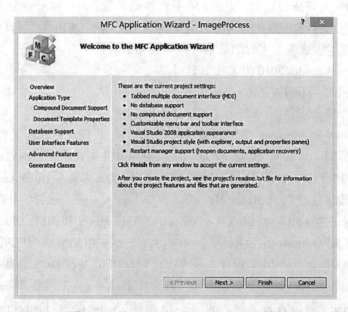

图 1-14　"MFC Application Wizard"对话框

图 1-15 所示的"Application Type"对话框，在 Visual C++6.0 基础上添加了"Project Style"等关于工程外观样式设计内容，可根据自己需要设计 MFC 界面。选择应用程序类型，我们看到有四种类型：Single document（单文档）、Multiple documents（多文档）、Dialog based（基于对话框）和 Multiple top-level documents。我们选择 Single document 类型，以生成一个单文档应用程序框架。单文档应用程序运行时是一个单窗口界面。

此对话框的"Resource language"还提供语言的选择，这里默认选择英语。"Project style"可选择工程风格，我们选择默认的"Visual Studio"风格。"Use of MFC"有两个选项：Use MFC in a shared DLL（动态链接库方式使用 MFC）和 Use MFC in a static library（静态库方式使用 MFC）。选择 Use MFC in a shared DLL 时，MFC 的类会以动态链接库的方式访问，所以我们的应用程序本身就会小些，但是发布应用程序时必须同时添加必要的动态链接库，以便在没有安装 VS2010 的电脑上能够正常运行程序。选择 Use MFC in a static library 时，MFC 的类会编译到可执行文件中，所以应用程序的可执行文件要比上种方式大，但可以单独发布,不需另加包含 MFC 类的库。这里我们使用默认的 Use MFC in a shared DLL。单击"Next"按钮。

图 1-15 "Application Type"对话框

图 1-16 所示的"Compound Document Support"对话框,与 Visual C++6.0 的"MFC AppWizard-Step 3"对话框内容相同,可以通过它向应用程序加入 OLE 支持,指定 OLE 选项的复合文档类型。

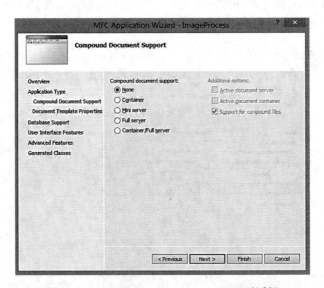

图 1-16 "Compound Document Support"对话框

图 1-17 所示的"Document Template Properties"对话框,与 Visual C++6.0"Document

Template Strings"对话框内容相近,可以设置程序能处理的文件的扩展名。对话框其他选项还可以更改程序窗口的标题。

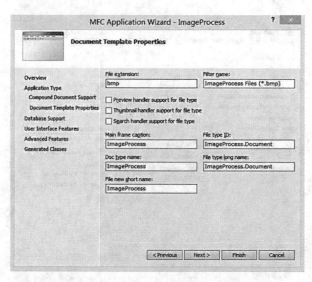

图 1-17 "Document Template Properties"对话框

图 1-18 所示的"Database Support"对话框,用于设置数据库选项。此向导可以生成数据库应用程序需要的代码。它有四个选项:None:忽略所有的数据库支持;Header files only:只包含定义了数据库类的头文件,但不生成对应特定表的数据库类或视图类;Database view without file support:创建对应指定表的一个数据库类和一个视图类,不附加标准文件支持;Database view with file support:创建对应指定表的一个数据库类和一个视图类,并附加标准文件支持。

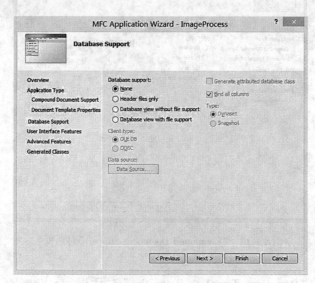

图 1-18 "Database Support"对话框

图 1-19 所示"User Interface Features"对话框,即用户界面特性。我们可以设置有无最大化按钮、最小化按钮、系统菜单和初始状态栏等。还可以选择使用菜单栏和工具栏生成简单的应用程序还是使用 ribbon。这里我们都选择默认设置。单击"Next"进入下一步。

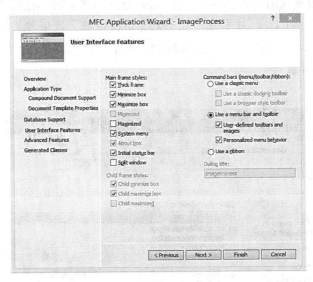

图 1-19 "User Interface Features"对话框

图 1-20 所示"Advanced features"对话框,可以设置的高级特性包括有无打印和打印预览等。在"Number of files on recent file list"项可以设置在程序界面的文件菜单下面最近打开文件的个数。我们仍使用默认值,单击"Next"按钮。

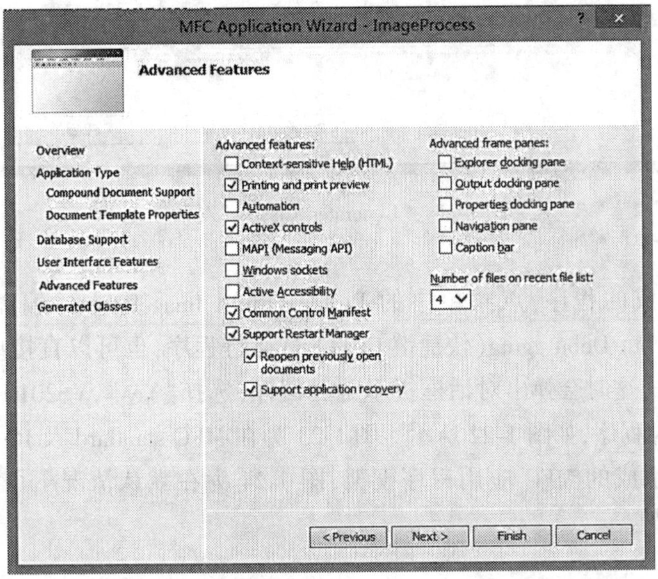

图 1-20 "Advanced features"对话框

图1-21所示"Generated Classes"对话框,在其上部的"Generated classed"列表框内,列出了将要生成的四个类:一个视图类(CImageProcessView)、一个应用类(CImageProcessApp)、一个文档类(CImageProcessDoc)和一个主框架窗口类(CMainFrame)。在对话框下面的几个编辑框中,可以修改默认的类名、类的头文件名和源文件名。对于视图类,还可以修改其基类名称,默认的基类是CView,还有其他几个基类可以选择。这里我们还是使用默认设置。单击"Finish"按钮。应用程序向导最后为我们生成了应用程序框架,并在Solution Explorer中自动打开了解决方案。

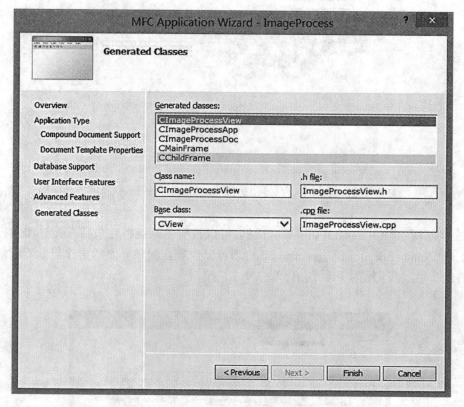

图1-21 "Generated Classes"对话框

编译运行生成的程序,点菜单中的 Build→Build ImageProcess 编译程序,然后单击 Debug→Start Without Debugging(快捷键 Ctrl+F5)运行程序,也可以直接单击 Debug→Start Without Debugging,这时会弹出对话框提示是否编译,选择"Yes",VS2010将自动编译链接运行 ImageProcess 程序,如图1-22所示。图1-23 为在 MFC standard 及 Resource language 设置为中文情况下生成的 MFC 应用程序框架,图1-24 为在默认情况下由应用向导生成的 MFC 应用程序框架。

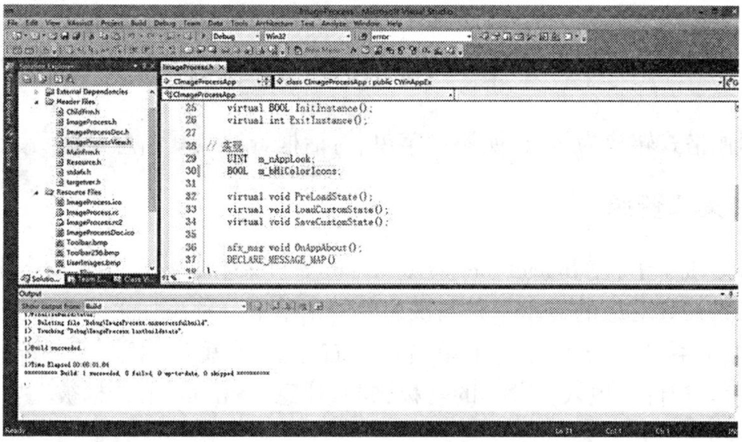

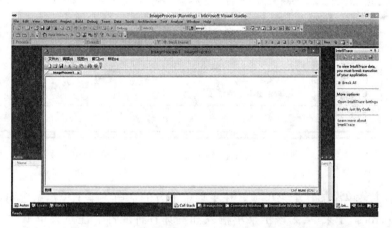

图 1-22　自动编译链接工程开发环境界面

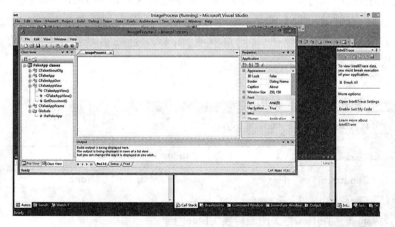

图 1-23　生成的 MFC standard 应用程序框架

图 1-24　在默认情况下由应用向导生成的 MFC 应用程序框架

第二节　应用程序中添加菜单、对话框资源及其消息处理函数的方法

下面以图像格式转换为例,说明添加菜单、对话框资源及其消息处理函数的方法。

一、添加相关菜单资源

在 ResourceView 中,在 ImageProcessEx Resource 目录下,双击"Menu"子目录下的"IDR_IMAGEPTYPE",在其对应的显示于右侧视窗中的菜单上分别添加主菜单及其子菜单项。

首先,添加主菜单,在空白菜单上单击鼠标右键,选择弹出菜单的"Properties"子菜单,在"Caption"编辑控件中填入主菜单的名称,同时注意"Pop-up"复选框被选中,如图 1-25 所示。

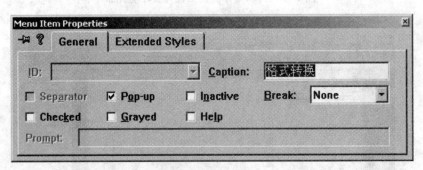

图 1-25　为工程添加"格式转换"主菜单项

然后,为主菜单添加其子菜单项。在主菜单下方的空白菜单上单击鼠标右键,选择弹出菜单的"Properties"子菜单,为该子菜单输入和选择各种属性。注意菜单 ID 号最好根据它所代表的内容来命名,如"Raw to Bmp"菜单项可取其 ID 为"ID_RAWTOBMP",如图 1-26 所示。菜单设计好后,可以进行一次编译连接并运行,观察 Raw to Bmp 菜单项是否已经出现。可见它是灰色的,表明处于没有激活状态。这是因为还没有为该菜单项添加消息处理函数。按照同样的方法,可以为工程添加其他菜单项,图 1-27 是按实习教程内容添加好所有菜单后显示的结果。

图 1-26　"格式转换"菜单下"Raw to Bmp"子菜单项的属性

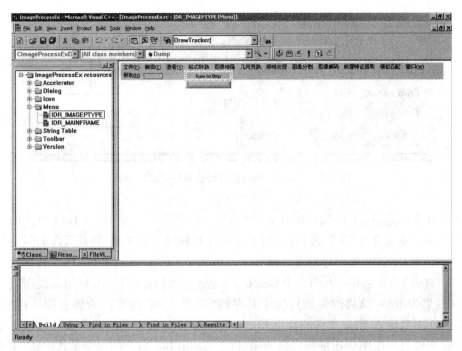

图 1-27　ImageProcessEx 工程添加完成的所有菜单资源

二、添加对话框资源

将 Raw 格式转换为 Bmp 格式时,一般需要给出文件路径、大小、颜色数等信息。因此,可以设计一个对话框来输入和设置此信息。添加对话框的过程为:

首先,在 ResourceView 中双击"Dialog",子目录上单击鼠标右键,在弹出的菜单中选择"Insert Dialog"子项,将自动出现一个默认对话框(图 1-28)。

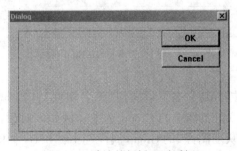

图 1-28　默认的新插入对话框

然后在对话框上单击鼠标右键,选择弹出菜单中的"Properties"子菜单为该对话框设置属性,如图 1-29 所示。

最后为对话框添加需要的控件。

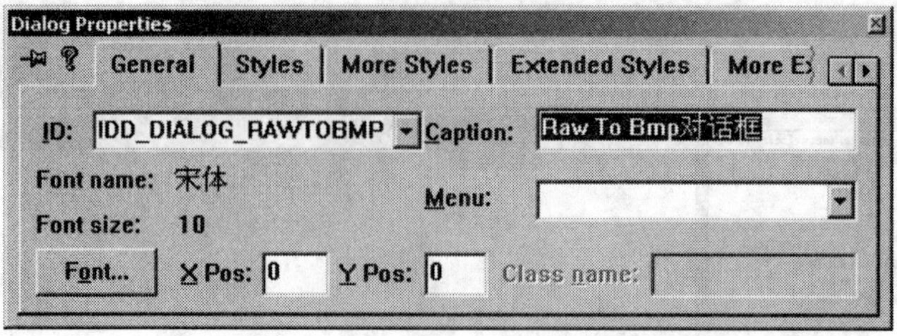

图 1-29　Raw To Bmp 格式转换对话框属性设置

图 1-30 所示是设计的"Raw To Bmp"对话框。需要说明的是：在图 1-30 中，"Raw 格式文件"和"Bmp 格式文件"为静态控件，用于说明打开和保存文件类型；其后面的两个 Edit 控件用于输入和显示打开和保存的文件路径、名称；两个"浏览"按钮控件，用于选择所要打开或保存文件的路径和名称；下面一个 Group Box 控件，类似静态控件，只用来说明其 Box 中的内容（这里为 RAW 文件的参数）；位于其左侧"宽度："和"高度："为静态控件；旁边的两个 Edit 控件用来输入 Raw 文件的高度和宽度；Group Box 的中间两个单选按钮确定所打开的 Raw 文件是灰度图像还是真彩色图像；位于 Group Box 右侧的三个单选按钮，用来选取 Raw 文件的存储方式：BSQ（逐波段存储）、BIL（逐行存储）和 BIP（逐像素存储）。

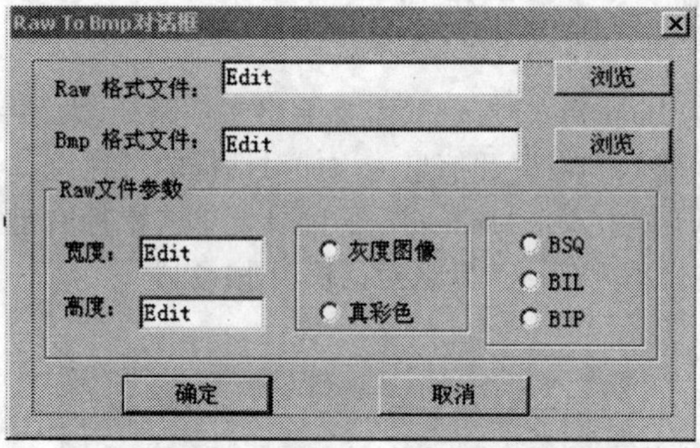

图 1-30　设计的"Raw To Bmp"对话框

注意：（1）给每个控件的 ID 号名尽量做到"见名知意"；（2）添加一组单选按钮时，要连续添加，并将第一个单选按钮属性中的 Group 选中。例如该对话框中的"灰度图像"和"真彩色"要连续添加，且将"灰度图像"的 Group 属性选中，"BSQ"、"BIL"和"BIP"也为一组，三个要连续添加，且将"BSQ"的 Group 属性选中。

三、为对话框创建类

在"Raw To Bmp 对话框"上点击鼠标右键，再点击弹出菜单的"ClassWizard"菜单项，将会在"MFC ClassWizard"对话框上弹出一个"Adding a Class"对话框，如图 1-31 所示。选中

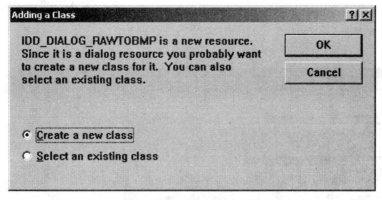

图 1-31 "Adding a Class"对话框

"Create a new class"项,点击"OK"按钮后,将弹出"New Class"对话框,如图 1-32 所示。注意对话框 ID 为 IDD_DIALOG_RAWTOBMP,即当前选中的对话框,基类为 CDialog,因为这是要创建一个关于对话框的类。在 Class information 的 Name 项中输入所要创建的类名,则 File name 中的内容自动与输入的类名一致。这里需要提醒的是,一般类名以字母 C 开头,同样也要做到见名知意。

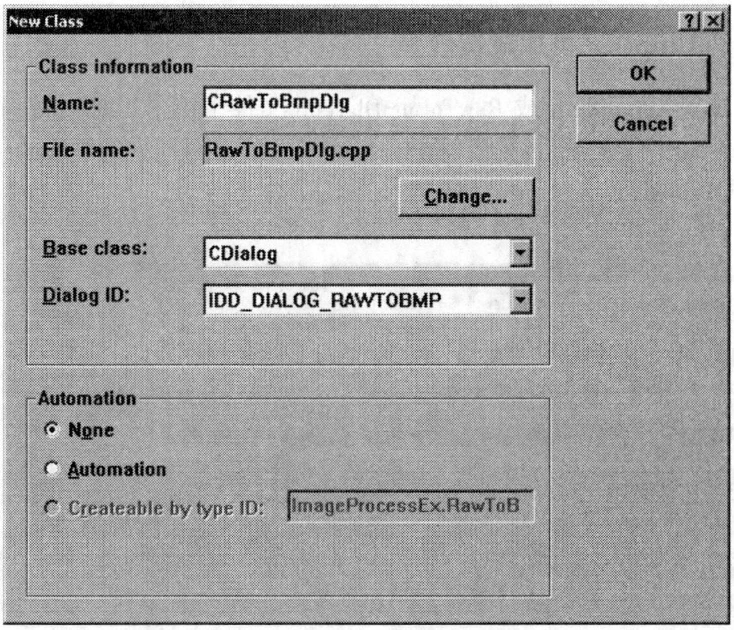

图 1-32 输入类名后的"New Class"对话框

类名输入完成后,单击"OK"按钮,系统自动创建新类 CRawToBmpDlg,伴随它的是 RawToBmpDlg.h 和 RawToBmpDlg.cpp,同时可以看到图 1-33 的"MFC ClassWizard"对话框,会发现刚才添加的新类已经在类列表中。单击 ClassView,会发现增加了新类 CRawTo

BmpDlg，双击 CRawToBmpDlg 会打开 RawToBmpDlg.h，即类定义文件（请注意仔细观察生成的头文件和源文件）。

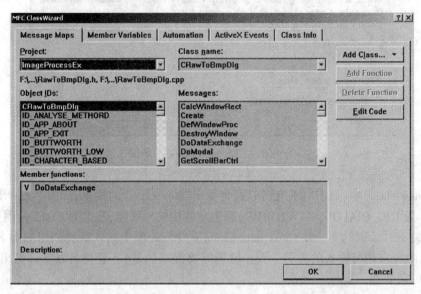

图 1-33　添加 CRawToBmpDlg 类后的"MFC ClassWizard"对话框

四、为 CRawToBmpDlg 添加成员变量

在打开的 RawToBmpDlg.h 或 RawToBmpDlg.cpp 文件中单击鼠标右键，在弹出的菜单中，单击"ClassWizard"菜单项，再单击"MFC ClassWizard"对话框中的"Member Variables"属性页，结果如图 1-34 所示。

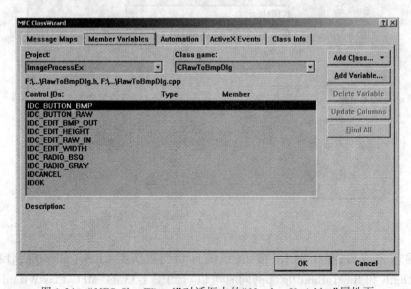

图 1-34　"MFC ClassWizard"对话框中的"Member Variables"属性页

根据需要选中需要添加变量的 ID，单击"Add Variable..."按钮，输入变量名，选择变量种类是数值还是控件。以 IDC_EDIT_BMP_OUT 为例，为了得到一个文件名字符串，选变量类型为 CString，同时，在 Catregory 中选择"Value"，如图 1-35 所示。检查无误后单击"OK"按钮。

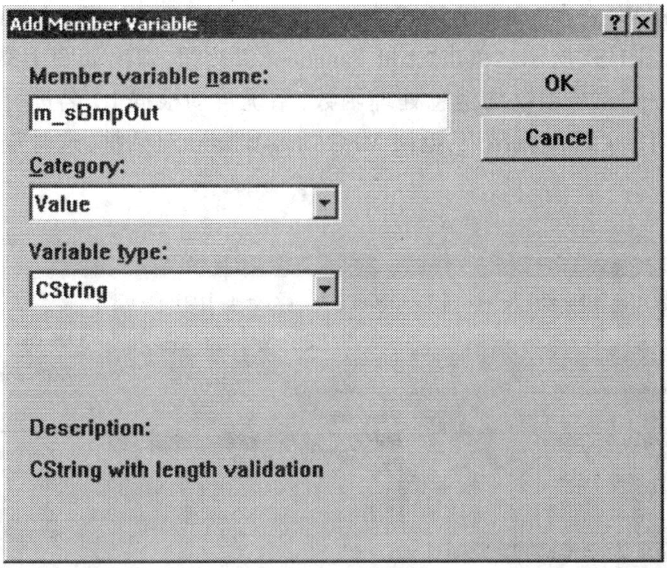

图 1-35　为 CRawToBmpDlg 的"IDC_EDIT_BMP_OUT"控件添加成员变量

按上述方法依次对不同 ID 添加相应变量，最后结果如图 1-36 所示。观察 CRawToBmpDlg、DlgRawToBmpDlg.h 或 RawToBmpDlg.cpp，注意其发生的变化。

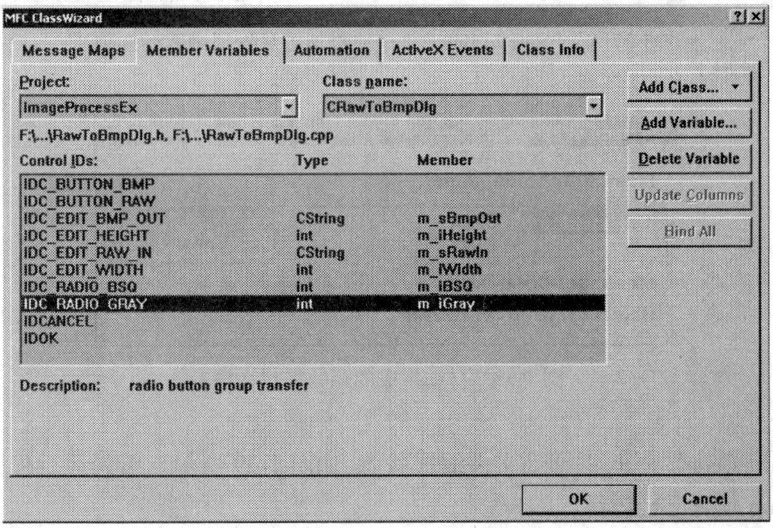

图 1-36　CRawToBmpDlg 添加的所有成员变量

五、添加消息处理函数

首先点击"ResourceView",接着双击 Menu 中的"IDR_IMAGEPTYPE"(不同工程该项的名称不同),选中前面添加的菜单项"Raw to Bmp"。在其上单击鼠标右键,在弹出的菜单上单击"ClassWizard"项,出现"MFC ClassWizard"对话框。再在对话框"ClassName"列表中选取 CImageProcessExView,在"Object IDs"中选中"ID_RAWTOBMP",在"Messages"中选中"COMMAND",如图 1-37 所示。单击"Add Function…"按钮,弹出如图 1-38 所示的菜单,这时可为"Raw to Bmp"菜单项添加消息处理函数,系统自动根据 ID 名进行函数命名。点击"OK"按钮,添加消息处理函数结束,返回 MFC ClassWizard 对话框,这时观察该对话框所发生的变化。

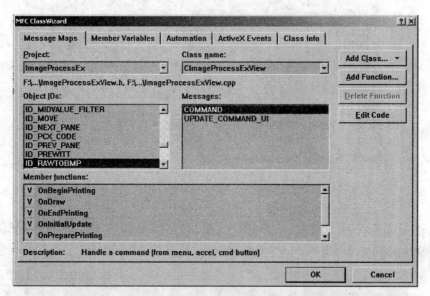

图 1-37 为"Raw to Bmp"菜单项添加消息处理函数的选项

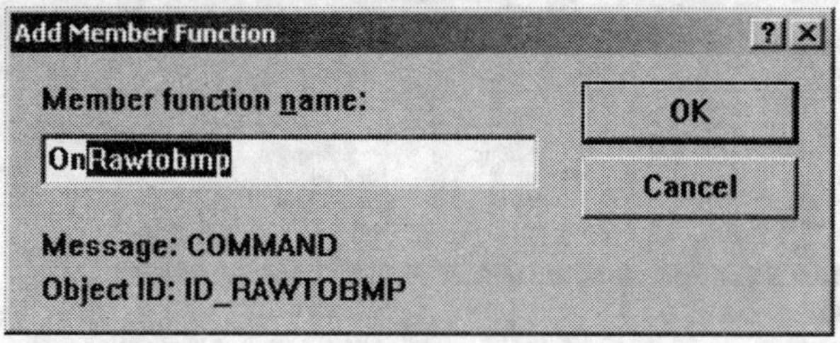

图 1-38 自动生成的函数名对话框

点击"Edit Code"按钮,返回到文件编辑状态,如图 1-39 所示。编译运行程序,观察此时的 Raw to Bmp 是否已经激活。

需要指出的是,添加菜单的消息处理函数可以在不同类中进行,如在 Doc、App、Frame

等类中均可,可根据编程便利情况来选择。

//
//CImageProcessExView message handlers
void CImageProcessExView::OnRawtoBmp()
{
　　//TODO:Add your command handler code here
}

图1-39　为"Raw to Bmp"添加的函数体

六、添加消息处理函数代码

在CImageProcessExView::OnRawtoBmp()的函数体中添加如下代码:
void CImageProcessExView::OnRawtoBmp()
{
　　// TODO:Add your command handler code here
　　CRawToBmpDlg dlg;　　　　　//定义格式转换对话框类对象
　　dlg.DoModal();　　　　　　//用来弹出格式转换对话框
}

进行一次编译,这时会提示四个出错信息。这些错误与刚刚添加的代码有关,错误原因是不识别 CRawToBmpDlg,其余错误都是由它所引起。修改该错误的方法就是在 ImageProcessExView.cpp 文件中添加头文件代码"RawToBmpDlg.h",如下面粗体字所示。
#include "stdafx.h"
#include "ImageProcessEx.h"

#include "ImageProcessExDoc.h"
#include "ImageProcessExView.h"
#include "RawToBmpDlg.h"

#ifdef _DEBUG
#define new DEBUG_NEW
#undef THIS_FILE
static char THIS_FILE[] = __FILE__;
#endif

这样就可以识别 CRawToBmpDlg 类型变量了。再次编译无误,连接并运行,运行结果如图1-40所示。可见"Raw to Bmp"菜单项激活且单击后可以弹出"Raw to Bmp"对话框。但此时的对话框中各个按钮没有任何响应,这是由于没有在 CRawToBmpDlg 中添加各响应函数。

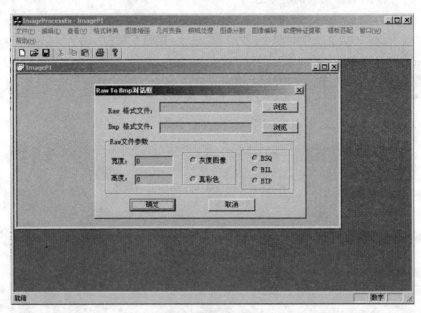

图 1-40　为"Raw to Bmp"菜单添加消息处理函数后运行的结果

七、为 CRawToBmpDlg 添加各消息处理函数

先回到 ResourceView，双击 Dialog 中的 IDD_DIALOG_RAWTOBMP，然后选中第一个浏览按钮，在其上单击鼠标右键，在弹出的菜单上选择"ClassWizard…"或"Events…"菜单项，为该浏览按钮添加消息处理函数，其结果分别如图 1-41 和图 1-42 所示。

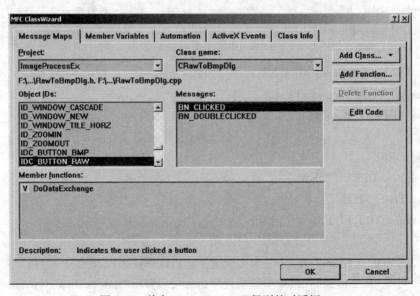

图 1-41　单击"ClassWizard…"得到的对话框

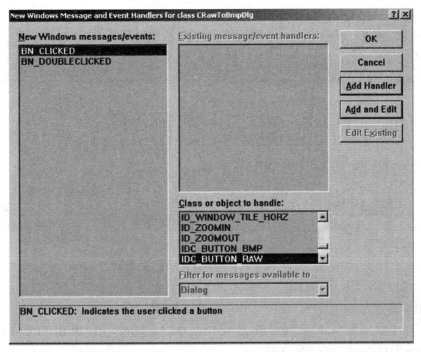

图 1-42 单击"Events…"得到的对话框

在图 1-41 或图 1-42 弹出的对话框中选中 BN_CLICKED 消息,单击 Add Function、Add Handler 或者 Add and Edit 按钮,系统会自动弹出函数名 OnButtonRaw()。若该函数名不用更改,则在弹出的对话框中单击 OK 按钮,系统就自动添加该函数的声明和函数体,这时在函数体内添加按钮被单击时需做出的响应。

下面是在 OnButtonRaw()函数体内添加代码的一个例子。

```
void CRawToBmpDlg::OnButtonRaw()
{
    //文件类型说明字符串
    static char BASED_CODE file[] = "RAW Files（*.RAW)|*.raw|所有文件（*.*）|*.*||";
    //文件对话框初始化,这里的第一个参数 TRUE 表示以打开方式显示文件对话框
    CFileDialog SelectFile（TRUE, NULL, NULL, OFN_HIDEREADONLY | OFN_OVERWRITEPROMPT, file, NULL）;
    //弹出文件打开对话框
    SelectFile.DoModal();
    //得到所选文件路径(包括文件名)
    CString FileName;
    FileName=SelectFile.GetPathName();
```

//将得到的文件名赋给格式转换对话框的成员变量 m_sRawIn

m_sRawIn=FileName;

//数据刷新,即将 m_sRawIn 新得到的值显示在对话框中对应的控件上(注意参数 FALSE)

UpdateData(FALSE);

}

按上述方法还可以给其他相关按钮添加消息相应函数。比如:

(1)文件保存为 BMP 路径选择按钮 BN_CLICKED 的消息处理函数。

void CRawToBmpDlg::OnButtonBmp()

{

//文件类型说明字符串,打开 BMP 文件

static char BASED_CODE file[ ] = "BMP Files(*.BMP)|*.BMP|所有文件(*.*)|*.*||";

//文件对话框初始化,这里的第一个参数 FALSE 表示以保存方式显示文件对话框

//第二个参数"BMP"用于指定默认保存文件的扩展名,即保存时自动添加扩展名 "BMP"

CfileDialog SelectFile(FALSE," BMP ", NULL, OFN_HIDEREADONLY | OFN_OVERWRITEPROMPT, file, NULL);

//弹出文件打开对话框

SelectFile.DoModal();

//得到所选文件路径(包括文件名)

CString FileName;

FileName=SelectFile.GetPathName();

//将得到的文件名赋给格式转换对话框的成员变量 m_sBmpOut

m_sBmpOut=FileName;

//数据刷新

UpdateData(FALSE);

}

(2)单击"确定"按钮的操作代码。

void CRawToBmpDlg::OnOK()

{

// TODO: Add extra validation here

//数据刷新

UpdateData(TRUE);

//声明 DIB 句柄 HDIB

DECLARE_HANDLE(HDIB);

HDIB hDIB;//定义 HDIB 句柄变量,用来存放 BMP 位图

LPSTR pDIB;//定义字符指针变量,用来存放位图数据

//打开文件
CFile fileraw,filebmp; //CFile 读写文件
//以读和拒绝写的方式打开所选文件名为 m_sRawIn 的 Raw 格式文件
fileraw.Open(m_sRawIn,CFile::modeRead | CFile::shareDenyWrite,NULL);
//得到文件的大小
DWORD filesize =fileraw.GetLength();
//根据 Raw 格式文件大小来申请分配内存空间
hDIB =(HDIB)::GlobalAlloc(GMEM_MOVEABLE| GMEM_ZEROINIT, filesize);
//若分配不成功,即 hDIB ==0,则程序返回
if (hDIB ==0)
{
    return;
}
//若分配成功,得到句柄内存的起始地址
pDIB =(LPSTR)::GlobalLock((HGLOBAL) hDIB);
//将 Raw 文件内容读到 pDIB 所指向的内存块
if(m_iGray==0)//如果 Raw 文件,则不需要考虑 m_iBSQ
    fileraw.ReadHuge(pDIB,filesize);
else //如果 Raw 文件是真彩色,需要考虑 m_iBSQ
{
    if(m_iBSQ==0)
    {
        //这里由学生添加
        AfxMessageBox("真彩色 BSQ 方式没有实现,由学生添加", MB_OK, 0);
        return;
    }
    if(m_iBSQ==1)
    {
        //这里由学生添加
        AfxMessageBox("真彩色 BIL 方式没有实现,由学生添加", MB_OK, 0);
        return;
    }
    if(m_iBSQ==2)
    {
        //这里由学生添加
        AfxMessageBox("真彩色 BIP 方式没有实现,由学生添加", MB_OK, 0);
        return;
    }

}
//不允许其他程序同时打开所选 BMP 文件
filebmp. Open (m _ sBmpOut, CFile：： modeCreate ｜ CFile：： modeWrite ｜ CFile：： shareExclusive,NULL）；

//写文件：文件头、位图信息头、调色板、位图数据
//根据 BMP 文件由四部分组成,下面分别给各部分赋值并写入 BMP 文件
BITMAPFILEHEADER bmfHdr；   // 位图文件头
BITMAPINFOHEADER bmpHdr；   // 位图信息头
RGBQUAD rgb[256]；   // 调色板

//位图文件头部分信息初始化
bmfHdr.bfType = ′MB′；
bmfHdr.bfReserved1 = 0；
bmfHdr.bfReserved2 = 0；

//位图信息头部分信息初始化
bmpHdr.biSize = 40；
bmpHdr.biWidth = m_iWidth；
bmpHdr.biHeight = m_iHeight；
bmpHdr.biPlanes = 1；
bmpHdr.biCompression = BI_RGB；
bmpHdr.biXPelsPerMeter = 0；
bmpHdr.biYPelsPerMeter = 0；
bmpHdr.biClrUsed = 0；
bmpHdr.biClrImportant = 0；

if(m_iGray = = 0)
{
　//若 Raw 文件为灰度图像,给相应位图文件头和位图信息头部分内容初始化
　bmpHdr.biBitCount = 8；
　bmfHdr.bfOffBits = 1078；//bmfHdr.bfOffBits = 14+40+256 * 4；
　bmpHdr.biSizeImage = (((m_iWidth * 8) + 31) ／ 32 * 4) * m_iHeight；
　bmfHdr.bfSize = bmfHdr.bfOffBits+bmpHdr.biSizeImage；
　//调色板初始化
　for(int i = 0；i<256；i++)
　{

```
            rgb[i].rgbBlue=i;
            rgb[i].rgbGreen=i;
            rgb[i].rgbRed=i;
            rgb[i].rgbReserved=0;
        }
    }
    else
    {
        //若 Raw 文件为真彩色图像,给相应位图文件头和位图信息头部分内容初始化
        bmpHdr.biBitCount=24;
        bmfHdr.bfOffBits=54;//bmfHdr.bfOffBits=14+40;
        bmpHdr.biSizeImage=((( m_iWidth*24)+31)/32*4)*m_iHeight;
        bmfHdr.bfSize=bmfHdr.bfOffBits+bmpHdr.biSizeImage;
    }

    //将初始化好的位图文件头和位图信息头写入文件(注意是按位图文件结构顺序来写的)
    filebmp.Write(&bmfHdr,sizeof(bmfHdr));//写位图文件头
    filebmp.Write(&bmpHdr,sizeof(bmpHdr));//写位图信息头
    if(m_iGray==0)
            filebmp.Write(rgb,sizeof(RGBQUAD)*256);//如果是灰度图像,将调色板写入文件

    int h=m_iHeight;
    int w=m_iWidth;
    int iWidthBytes=bmpHdr.biSizeImage/m_iHeight;//位图文件的实际每行的存储宽度
    //开始写第四部分(图像数据)
    //注意每一行的字节数必须是四的整数倍,不足的补零
    //像素按照从下到上、从左到右的顺序排列
    int zero=0;
    for(int i=0;i<h;i++)
    {
        filebmp.Write(pDIB+w*(h-i-1),w);
        if(w!=iWidthBytes)
            for( int j=0;j<iWidthBytes-w;j++)
                filebmp.Write(&zero,1);
```

}

```
//释放句柄
::GlobalUnlock((HGLOBAL)hDIB);
::GlobalFree((HGLOBAL)hDIB);
//关闭文件
fileraw.Close();
filebmp.Close();

CDialog::OnOK();
}
```

通过添加以上菜单、对话框资源及其相应的消息处理函数,编译运行,就能实现 Raw 格式文件到 BMP 文件的格式转换功能。

第二章 位图操作 CDib 类的实现

第一节 位图文件格式与调色板原理

位图是采用位映像方法存储的图像。位图有很多种格式,这里仅提及与实习有关的 Raw 和 DIB 两种。

一、Raw 文件

Raw 文件将像素按行列号顺序存储在文件中。这种文件只含有图像像素数据,不含有信息头,因此,在读图像时,需要事先知道图像大小(行列数)、颜色数及其存储方式(如 BSQ、BIL、BIP 方式)。它是最简单的一种图像文件格式。

二、DIB 文件

DIB 文件也称设备无关位图(Device-Independent-Bitmap,DIB),它是标准的 Windows 位图格式,它通常以 BMP 文件格式保存。关于 BMP 文件格式在教材中已作介绍,这里仅详细介绍调色板的原理。

位图是由一个个像素组成的,每个像素都有自己的颜色属性。像素的颜色是基于 RGB 模型的,每一个像素的颜色由红(R)、绿(G)、蓝(B)三原色组合而成。若每种原色用 8 位表示,这样一个像素的颜色就是 24 位的。24 位的颜色通常被称做真彩色,用真彩色显示的图像可达到十分逼真的效果。

但是,真彩色的显示需要大量的内存,一幅 640×480 的真彩色图像需要约 1MB 的内存。由于彩色图像数据量大,显示真彩色会使系统的整体性能迅速下降。为了解决这个问题,计算机图像显示采用调色板来限制颜色的数目。调色板实际上是一个有 256 个表项的 RGB 颜色表,颜色表的每项是一个 24 位的 RGB 颜色值。使用调色板时,在内存中存储的不是 24 位颜色值,而是调色板的 4 位或 8 位的索引。这样,显示器同时显示的颜色被限制在 256 色以内,但对系统资源的耗费大大降低了。

位图文件中的颜色列表(调色板)设置取决于图像是索引图像还是真彩色图像。如果图像是索引色图像,则位图文件中设有调色板;如果图像是真彩色图像,则位图文件中没有调色板。使用调色板的一个好处是不必改变内存中的值,只需改变调色板的颜色项就可快速地改变一幅图像的颜色或灰度。

显示模式设置有 16K、256K、64K、真彩色等,前两种模式需要调色板。在 16 色或 256 色模式下,读取索引色图像时,根据图像每个像素的索引值(灰度值)在调色板中找到它们相应的颜色,来获得各像素的颜色,然后将索引色图像显示出来。图 2-1 给出了调色板的工作原理。

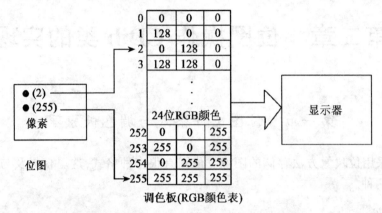

图 2-1 调色板工作原理

Windows 是一个多任务操作系统,可以同时运行多个程序。如果有几个程序都要设置调色板,就有可能产生冲突。为了避免这种冲突,Windows 使用逻辑调色板作为使用颜色的应用程序和系统调色板(物理调色板)之间的缓冲。

在 Windows 中,应用程序是通过一个或多个逻辑调色板来使用系统调色板(物理调色板)。在 256 色系统调色板中,Windows 保留了 20 种颜色作为静态颜色,这些颜色用做显示 Windows 界面,应用程序一般不能改变。缺省的系统调色板只包含这 20 种静态颜色,调色板的其他项为空。应用程序要想使用新的颜色,必须将包含有所需颜色的逻辑调色板实现到系统调色板中。在实现过程中,Windows 首先将逻辑调色板中的项与系统调色板中的项作完全匹配,对于逻辑调色板中不能完全匹配的项,Windows 将其加入到系统调色板的空白项中,系统调色板总共有 236 个空白项可供使用,若系统调色板已满,则 Windows 将逻辑调色板的剩余项匹配到系统调色板中尽可能接近的颜色上。

灰度图像可看做只含亮度信息、不含色彩信息的图像。因此,要显示灰度图像,只需要将亮度值进行量化,通常划分成 0~255 共 256 个级别,0 表示最暗(全黑),255 表示最亮(全白)。然后利用 256 色的调色板实现图像显示。只不过这种调色板比较特殊,它的 R、G、B 分量每一项都相同,即 RGB 从 RGB(0,0,0)变到 RGB(255,255,255)。每一个像素的灰度值就是它在该调色板中的索引值。

第二节 位图操作类 CDib 的实现

为了实现对位图的操作,并将所有对位图的操作都封装在一个类中,方便继承和移植,充分体现面向对象程序 OOP 设计的优越性,这里介绍类、函数和位图操作类 Cdib 的实现。

一、有关文件操作类

CFileDialog：用于实现打开/保存文件对话框。
CFile：用于实现文件的打开、读写以及定位等功能。

二、内存操作函数

HGLOBAL GlobalAlloc(UINT uFlags, // allocation attributes
    DWORD dwBytes // number of bytes to allocate
    );分配指定类型和大小的内存，若分配成功，返回所分配的内存句柄；
GlobalLock(HGLOBAL hMem)：锁定全局内存对象并返回该内存块的起始地址；
GlobalUnLock（HGLOBAL hMem）：减少该内存对象申请块数；
GlobalFree（HGLOBAL hMem）：释放该内存对象并使句柄无效。

三、位图显示函数

（1）int SetDIBitsToDevice(
  HDC *hdc*,   //设备上下文句柄，它可以是 CDC 对象的公共成员变量 m_hDC
  int *XDest*,   //指定绘图区域的左上角 x 坐标（逻辑单位）
  int *YDest*,   //指定绘图区域的左上角 y 坐标（逻辑单位）
  DWORD *dwWidth*,   //指定位图的宽度（逻辑单位）
  DWORD *dwHeight*,   //指定位图的高度（逻辑坐标）
  int *XSrc*,   //指定数据源区域的左下角 x 坐标（逻辑单位）
  int *YSrc*,   //指定数据源区域的左下角 y 坐标（逻辑单位）
  UINT *uStartScan*,   //指定位图扫描的起始行
  UINT *cScanLines*,   //指定位图扫描的行数（即位图的高度）
  CONST VOID **lpvBits*,   //指向位图数据的指针
  CONST BITMAPINFO **lpbmi*,   //指向 BITMAPINFO 结构的指针
  UINT *fuColorUse*   //指定是 RGB 值还是调色板索引值
);

该函数可以直接在显示器或打印机上显示位图，在显示时不进行缩放处理，即位图的每一个像素对应于一个显示像素或打印机的打印点。

（2）int StretchDIBits(
  HDC *hdc*,   //设备上下文句柄
  int *XDest*,   //指定绘图区域的左上角 x 坐标（逻辑单位）
  int *YDest*,   //指定绘图区域的左上角 y 坐标（逻辑单位）
  int *nDestWidth*,   //指定绘图区域的宽度（逻辑单位）

```
    int nDestHeight,              //指定绘图区域的高度(逻辑单位)
    int XSrc,                     //指定数据源区域的左上角 x 坐标(逻辑单
    位)
    int YSrc,                     //指定数据源区域的左上角 y 坐标(逻辑单
    位)
    int nSrcWidth,                //数据源区域的宽度(逻辑单位)
    int nSrcHeight,               //数据源区域的高度(逻辑单位)
    CONST VOID * lpBits,          //指向位图数据区的指针
    CONST BITMAPINFO * lpBitsInfo, //指向 BITMAPINFO 结构的指针
    UINT iUsage,                  //指定是 RGB 值还是调色板索引值
    DWORD dwRop                   //指定绘制方式
);
```

该函数可以实现对位图的放大和缩小显示。

四、位图操作类 CDib 的定义和实现

为了实现对位图的操作,并将所有对位图的操作都封装在一个类中,方便继承和移植,充分体现面向对象程序 OOP 设计的优越性,这里给大家介绍一个位图操作类 CDib 的定义和实现。

CDib 类的头文件和源文件分别为:dib.h、dib.cpp。其中 dib.h 用于实现类的定义,dib.cpp 用于实现类的各种操作。实际上该类只是对上述 API 函数功能的封装,即所有函数及变量都由 CDib 类来实现。这样就可以将所有位图操作的变量和函数在该类中进行定义和实现。下面介绍这一示例,要求能掌握类的概念和应用,吸收 OOP 思想,从而用面向对象的方法实现各种操作。

1. CDib 类的定义

```
// dib.h

#ifndef _INC_DIB
#define _INC_DIB

/* 位图常数 */
#define PALVERSION      0x300

/* 位图头文件标记 */
#define DIB_HEADER_MARKER    ((WORD) ('M' << 8) | 'B')

/* 常用宏 */
#define RECTWIDTH(lpRect)      ((lpRect)->right - (lpRect)->left)
```

```
#define RECTHEIGHT(lpRect)      ((lpRect)->bottom - (lpRect)->top)
```

// WIDTHBYTES 表示图像扫描行宽

```
#define WIDTHBYTES(bits)        (((bits) + 31) / 32 * 4)

/* CDib 类定义 */
class CDib : public CObject
{
    DECLARE_DYNAMIC(CDib)

public:
    CDib();

protected:
    LPBYTE m_pBits;
    LPBITMAPINFO m_pBMI;
public:
    CPalette * m_pPalette;


public:
    DWORD Width()      const; //获取位图宽度
    DWORD Height()     const; //获取位图高度
    WORD  NumColors()  const; //返回位图的调色板索引位数
    BOOL  IsValid()    const { return (m_pBMI != NULL); }

public:
    BOOL   Paint(HDC, LPRECT, LPRECT) const; //显示位图
    HGLOBAL CopyToHandle()             const; //拷贝到句柄
    DWORD  Save(CFile& file)           const; //保存位图
    DWORD  Read(CFile& file); /读取位图文件
    DWORD  ReadFromHandle(HGLOBAL hGlobal);
    LPBYTE RETURN()
    {
        return m_pBits;
    }
    LPBYTE DD(LPBYTE m)
    {
```

```cpp
        m_pBits = m;
        return m_pBits;
    }
    void Invalidate( ) { Free( ); }

    virtual void Serialize( CArchive& ar );

// 执行部分
public:
    virtual ~CDib( );

protected:
    BOOL    CreatePalette( );
    WORD    PaletteSize( ) const;
    void    Free( );

public:
#ifdef _DEBUG
    virtual void Dump( CDumpContext& dc ) const;
#endif

protected:
    CDib& operator = ( CDib& dib );
};

#endif //! _INC_DIB
```

2.CDib 类的实现:
```cpp
// dib.cpp

#include "stdafx.h"
#include "dib.h"
#include <windowsx.h>
#include <afxadv.h>
#include <io.h>
#include <errno.h>
```

///

```cpp
// CDib

IMPLEMENT_DYNAMIC(CDib, CObject)
//构造函数,初始化
CDib::CDib()
{
    m_pBMI = NULL;
    m_pBits = NULL;
    m_pPalette = NULL;
}
//析构函数,释放内存
CDib::~CDib()
{
    Free();
}

void CDib::Free()
{
    // 确保所有将被分配内存的成员变量被释放
    if(m_pBits)
    {
        GlobalFreePtr(m_pBits);
        m_pBits = NULL;
    }
    if(m_pBMI)
    {
        GlobalFreePtr(m_pBMI);
        m_pBMI = NULL;
    }
    if(m_pPalette)
    {
        m_pPalette->DeleteObject();
        delete m_pPalette;
        m_pPalette = NULL;
    }
}

/***************************************************************
 * Paint()
```

```
    *  参数：
        * HDC hDC              ——设备环境句柄
        * LPRECT lpDCRect      ——设备环境矩形区域
        * LPRECT lpDIBRect     ——位图矩形区域
        * CPalette * pPal      ——位图调色板指针
    *  返回值：
        * BOOL                 ——绘制成功返回 TRUE,否则返回 FALSE
 *  说明：该函数主要功能是将位图从矩形区域 lpDIBRect,拷贝到设备环境 hDC 上的矩形
     区域 lpDCRect。
 *********************************************************** /

BOOL CDib::Paint(HDC hDC, LPRECT lpDCRect, LPRECT lpDIBRect) const
{
    if (! m_pBMI)
        return FALSE;

    HPALETTE hPal = NULL;            //新调色板句柄
    HPALETTE hOldPal = NULL;         //初始调色板

    // 获取位图的调色板
    if (m_pPalette != NULL)
    {
        hPal = (HPALETTE) m_pPalette->m_hObject;

        // 记录初始调色板
        hOldPal = ::SelectPalette(hDC, hPal, TRUE);
    }


    ::SetStretchBltMode(hDC, COLORONCOLOR);

    BOOL bSuccess;
    if (((RECTWIDTH(lpDCRect)  == RECTWIDTH(lpDIBRect)) &&
         (RECTHEIGHT(lpDCRect) == RECTHEIGHT(lpDIBRect))))
        bSuccess = ::SetDIBitsToDevice(hDC,              //设备环境句柄
                                    lpDCRect->left,      //绘制目标区域的左上角 x
                                                         //坐标
                                    lpDCRect->top,       //绘制目标区域的左上角 y
                                                         //坐标
```

```
                        RECTWIDTH(lpDCRect),    //绘制目标区域的宽度
                        RECTHEIGHT(lpDCRect),   //绘制目标区域的高度
                        lpDIBRect->left,        //位图区域的左上角 x 坐标
                        (int)Height() -
                        lpDIBRect->top -
                        RECTHEIGHT(lpDIBRect),  //位图区域左上角
                                                            y 坐标
                        0,                      // 开始扫描行
                        (WORD)Height(),         // 扫描总行数
                        m_pBits,                //位图数据头
                        m_pBMI,                 //位图信息头
                        DIB_RGB_COLORS);        //颜色参数
    else
        bSuccess = ::StretchDIBits(hDC,
                        lpDCRect->left,
                        lpDCRect->top,
                        RECTWIDTH(lpDCRect),
                        RECTHEIGHT(lpDCRect),
                        lpDIBRect->left,
                        lpDIBRect->top,
                        RECTWIDTH(lpDIBRect),
                        RECTHEIGHT(lpDIBRect),
                        m_pBits,
                        m_pBMI,
                        DIB_RGB_COLORS,
                        SRCCOPY);
    //恢复以前的调色板
    if (hOldPal != NULL)
    {
        ::SelectPalette(hDC, hOldPal, TRUE);
    }

    return bSuccess;
}
/***********************************************************
 * CreatePalette()
 * 返回值：BOOL——成功返回 TRUE，否则返回 FALSE
 * 说明：
 * 这个函数用来生成位图调色板
```

```cpp
* *************************************************** /
BOOL CDib::CreatePalette( )
{
    if (! m_pBMI)
        return FALSE;

    //获得颜色数
    WORD wNumColors = NumColors( );

    if (wNumColors != 0)
    {
        // 分配逻辑调色板空间
        HANDLE hLogPal = ::GlobalAlloc(GHND, sizeof(LOGPALETTE) + sizeof
(PALETTEENTRY) * wNumColors);

        // 内存不足,返回 FALSE
        if (hLogPal == 0)
            return FALSE;

        LPLOGPALETTE lpPal = (LPLOGPALETTE)::GlobalLock((HGLOBAL)hLogPal);

        //设置调色板
        lpPal->palVersion = PALVERSION;
        lpPal->palNumEntries = (WORD)wNumColors;

        for (int i = 0; i < (int)wNumColors; i++)
        {
            lpPal->palPalEntry[i].peRed = m_pBMI->bmiColors[i].rgbRed;
            lpPal->palPalEntry[i].peGreen = m_pBMI->bmiColors[i].rgbGreen;
            lpPal->palPalEntry[i].peBlue = m_pBMI->bmiColors[i].rgbBlue;
            lpPal->palPalEntry[i].peFlags = 0;
        }

        /* 生成调色板,并获得指针 */
        if (m_pPalette)
        {
            m_pPalette->DeleteObject( );
            delete m_pPalette;
        }
```

```
        m_pPalette = new CPalette;
        BOOL bResult = m_pPalette->CreatePalette(lpPal);
        ::GlobalUnlock((HGLOBAL)hLogPal);
        ::GlobalFree((HGLOBAL)hLogPal);
        return bResult;
    }


    return TRUE;
}


/***************************************************************
 * Width()
 * 返回值:
   * DWORD              ——位图的宽度
 * 说明:
 * 该函数获得位图的宽度
 ***************************************************************/
DWORD CDib::Width() const
{
    if(! m_pBMI)
        return 0;

    /* 返回宽度值 */
    return m_pBMI->bmiHeader.biWidth;
}
/***************************************************************
 * Height()
 * 返回值:
   * DWORD              ——位图的高度
 * 说明:
 * 函数返回图像的高度
 ***************************************************************/
DWORD CDib::Height() const
{
    if(! m_pBMI)
        return 0;

    /*返回高度值 */
    return m_pBMI->bmiHeader.biHeight;
```

```
/ ***************************************************************
 * PaletteSize()
 * 返回值：
 *   WORD               ——位图调色板大小
 * 说明：
 *   该函数获得位图调色板尺寸
 *************************************************************** /
WORD CDib::PaletteSize() const
{
    if (! m_pBMI)
        return 0;

    return NumColors() * sizeof(RGBQUAD);
}


/ ***************************************************************
 * NumColors()
 * 返回值：
 *   WORD               ——颜色数
 * 说明：
 *   这个函数获得颜色数
 *   如果位数为 1：colors=2，如果位数为 4：colors=16，如果位数为 8：colors=256,
 *   如果位数为 24,颜色表中没有颜色
 *************************************************************** /
WORD CDib::NumColors() const
{
    if (! m_pBMI)
        return 0;

    WORD wBitCount;    // DIB 的 bit 数

    /*  颜色表中的颜色数可以比像素允许的比特数少（例如 lpbi->biClrUsed
     *  可以设置为某一个值,根据该值返回合适的值） */

    DWORD dwClrUsed;

    dwClrUsed = m_pBMI->bmiHeader.biClrUsed;
```

```
    if (dwClrUsed != 0)
        return (WORD)dwClrUsed;

    /* 根据位图调色板计算颜色数 */
    wBitCount = m_pBMI->bmiHeader.biBitCount;

    /* 返回颜色数 */
    switch (wBitCount)
    {
        case 1:
            return 2;

        case 4:
            return 16;

        case 8:
            return 256;

        default:
            return 0;
    }
}

/****************************************************************
 * Save()
 * 参数:
 *   * CFile& file ——保存位图的文件
 * 返回值:
 *   * 存储字节数
 * 说明:
 *   * 将图像保存到指定文件
 ****************************************************************/
DWORD CDib::Save(CFile& file) const
{
    BITMAPFILEHEADER bmfHdr; // 位图文件头指针
    DWORD dwDIBSize;

    if (m_pBMI == NULL)
        return 0;
```

```
// 写文件头

// 文件类型最初 2bits 要是 "BM"
bmfHdr.bfType = DIB_HEADER_MARKER;    // "BM"

// 信息头和调色板大小
dwDIBSize = *(LPDWORD)&m_pBMI->bmiHeader + PaletteSize();

// 计算图像大小
if ((m_pBMI->bmiHeader.biCompression == BI_RLE8) || (m_pBMI->bmiHeader.biCompression == BI_RLE4))
{
    // 如果是 RLE 位图,不能计算图像尺寸,直接采用 biSizeImage
    dwDIBSize += m_pBMI->bmiHeader.biSizeImage;
}
else
{
    DWORD dwBmBitsSize;    // 位图数据块大小

    // 不是 RLE 位图, 大小为 Width (DWORD aligned) * Height
    dwBmBitsSize = WIDTHBYTES((m_pBMI->bmiHeader.biWidth) * ((DWORD) m_pBMI->bmiHeader.biBitCount)) * m_pBMI->bmiHeader.biHeight;
    dwDIBSize += dwBmBitsSize;

    m_pBMI->bmiHeader.biSizeImage = dwBmBitsSize;
}

// 计算整个文件大小,并书写图像文件头
bmfHdr.bfSize = dwDIBSize + sizeof(BITMAPFILEHEADER);
bmfHdr.bfReserved1 = 0;
bmfHdr.bfReserved2 = 0;

//计算图像文件头到图像数字的偏移
bmfHdr.bfOffBits = (DWORD)sizeof(BITMAPFILEHEADER) + m_pBMI->bmiHeader.biSize + PaletteSize();

//文件头
```

```cpp
    file.Write((LPSTR)&bmfHdr, sizeof(BITMAPFILEHEADER));
    DWORD dwBytesSaved = sizeof(BITMAPFILEHEADER);

    // 信息头
    UINT nCount = sizeof(BITMAPINFO) + (NumColors()-1) * sizeof(RGBQUAD);
    dwBytesSaved += nCount;
    file.Write(m_pBMI, nCount);

    // 数据部分
    DWORD dwBytes = m_pBMI->bmiHeader.biBitCount * Width();

    // 计算每行像素数据
    if (dwBytes%32 == 0)
        dwBytes /= 8;
    else
        dwBytes = dwBytes/8 + (32-dwBytes%32)/8 + (((32-dwBytes%32)%8 > 0) ? 1 : 0);

    nCount = dwBytes * Height();
    dwBytesSaved += nCount;
    file.WriteHuge(m_pBits, nCount);

    return dwBytesSaved;
}

/***************************************************************
 * 函数： Read(CFile&)
 * 返回值： 读入的 byte 数
 * 说明：将指定的位图文件读入内存
 ***************************************************************/

DWORD CDib::Read(CFile& file)
{
    // 释放内存
    Free();

    BITMAPFILEHEADER bmfHeader;

    // 判断文件是否存在
    if (file.Read((LPSTR)&bmfHeader, sizeof(bmfHeader)) != sizeof(bmfHeader))
```

```
        return 0;
    if(bmfHeader.bfType != DIB_HEADER_MARKER)
        return 0;
    DWORD dwReadBytes = sizeof(bmfHeader);

    // 分配内存
    m_pBMI = (LPBITMAPINFO) GlobalAllocPtr (GHND, bmfHeader.bfOffBits - sizeof
(BITMAPFILEHEADER) + 256 * sizeof(RGBQUAD));
    if(m_pBMI == 0)
        return 0;

    // 读文件头
    if(file.Read(m_pBMI, bmfHeader.bfOffBits-sizeof(BITMAPFILEHEADER)) != (UINT)
(bmfHeader.bfOffBits-sizeof(BITMAPFILEHEADER)))
    {
        GlobalFreePtr(m_pBMI);
        m_pBMI = NULL;
        return 0;
    }
    dwReadBytes += bmfHeader.bfOffBits-sizeof(BITMAPFILEHEADER);

    DWORD dwLength = file.GetLength();
    //读数据块
    m_pBits = (LPBYTE)GlobalAllocPtr(GHND, dwLength - bmfHeader.bfOffBits);
    if(m_pBits == 0)
    {
        GlobalFreePtr(m_pBMI);
        m_pBMI = NULL;
        return 0;
    }

    if(file.ReadHuge(m_pBits, dwLength-bmfHeader.bfOffBits) != (dwLength - bmfHeader.
bfOffBits))
    {
        GlobalFreePtr(m_pBMI);
        m_pBMI = NULL;
        GlobalFreePtr(m_pBits);
        m_pBits = NULL;
        return 0;
```

```
        dwReadBytes += dwLength - bmfHeader.bfOffBits;

    CreatePalette();
    return dwReadBytes;
}


#ifdef _DEBUG
void CDib::Dump(CDumpContext& dc) const
{
    CObject::Dump(dc);
}
#endif


/ **************************************************************
* 函数:     CopyToHandle
* 返回值：  指向新的全局存储空间
* 说明：    拷贝位图到一个全局存储空间
***************************************************************
HGLOBAL CDib::CopyToHandle() const
{
    CSharedFile file;
    try
    {
        if (Save(file) == 0)
            return 0;
    }
    catch (CFileException * e)
    {
        e->Delete();
        return 0;
    }

    return file.Detach();
}
/ **************************************************************
* 函数:     ReadFromHandle(HGLOBAL hGlobal)
* 返回值:读入的 Byte 数
* 说明：    初始化给定的内存空间
```

```
 ***************************************************************/
DWORD CDib::ReadFromHandle(HGLOBAL hGlobal)
{
    CSharedFile file;
    file.SetHandle(hGlobal, FALSE);
    DWORD dwResult = Read(file);
    file.Detach();
    return dwResult;
}
/////////////////////////////////////////////////////////////////////
/////////////////////// Serialization support

void CDib::Serialize(CArchive& ar)
{
    CFile * pFile = ar.GetFile();
    ASSERT(pFile != NULL);
    if (ar.IsStoring())
    {   // 存储函数
        Save(*pFile);
    }
    else
    {   // 载入数据
        Read(*pFile);
    }
}
```

第三章 基于 CDib 类的位图文件读取、显示和存储

在 VC 中，Doc 类是用来管理文件的。如果在 Doc 类中添加 CDib 类型的成员变量，也就是 CDib 的一个对象，那么所有对位图文件的操作就可通过 Doc 类的 CDib 类型的对象实现。因而所有对位图操作的修改和添加，只需在 CDib 类中进行就可以了。也就是 CDib 类实现了对位图操作的封装。这样不仅使用起来简便，而且移植也方便。只要将其对应的.h 文件和.cpp 文件拷贝到其他工程，那么其他工程就可以使用该类的所有功能了。

下面介绍利用 CDib 类实现位图文件读取、显示和存储的功能。其他操作的实现方法和步骤与此类似，按照下述方法添加其他操作，就可以不断丰富 CDib 类。

首先将 dib.h 和 dib.cpp 拷贝到所建的工程下，点击"Project/Add To Project/Files"菜单项，在弹出的对话框中选择 dib.h 和 dib.cpp 文件，点击"OK"按钮，这时会发现在 ClassView 视窗中增加了一个新类即 CDib 类。然后使用 CDib 类实现位图文件的读取、显示和存储。

第一节 位图读取

在 CImageProcessExDoc 中定义 CDib 类型的对象。在 ClassView 视窗中的 CImageProcessExDoc 类上单击右键，选择"Add Member Variable…"菜单项，在弹出的对话框中输入变量类型 CDib，变量名为 m_DIB 的 public 型变量，单击"OK"按钮后返回。这时会发现系统自动在 ImageProcessExDoc.h 文件中加入了下面两行代码：

#include "DIB.H" // 添加头文件
CDib m_DIB;

在 ClassView 的 CImageProcessExDoc 类上点击鼠标右键，在弹出的菜单上选择"Add Virual Function…"，选取 OnOpenDocument，再单击"Add and Edit"按钮，系统将自动回到添加的 OnOpenDocument( ) 函数的定义处，在 CImageProcessExDoc 类中添加虚函数 OnOpenDocument( )。在此函数体内添加如下用于打开位图文件的带底纹代码：

```
BOOL CImageProcessExDoc::OnOpenDocument(LPCTSTR lpszPathName)
{
    if (! CDocument::OnOpenDocument(lpszPathName))
        return FALSE;

    CFile file;
    CFileException fe;
```

```
if (! file.Open(lpszPathName, CFile::modeRead | CFile::shareDenyWrite, &fe))
{
    ReportSaveLoadException(lpszPathName, &fe, FALSE,
        AFX_IDP_FAILED_TO_OPEN_DOC);
    return FALSE;
}

DeleteContents();
BeginWaitCursor();

// 调用 Read DIBFile 函数读取位图
TRY
{
    m_DIB.Read(file);
}
CATCH (CFileException, eLoad)
{
    file.Abort();
    EndWaitCursor();
    ReportSaveLoadException(lpszPathName, eLoad, FALSE,
    AFX_IDP_FAILED_TO_OPEN_DOC);
    return FALSE;
}
END_CATCH

EndWaitCursor();

if (! m_DIB.IsValid())
{
    // may not be DIB format
    CString strMsg = "File can't open!";

    MessageBox(NULL, strMsg, NULL, MB_ICONINFORMATION | MB_OK);
    return FALSE;
}
SetPathName(lpszPathName);
SetModifiedFlag(FALSE);
return TRUE;
```

}

通过添加上述函数和代码,当执行程序的"文件/打开"菜单项时,若选择的文件是 DIB 位图,则位图信息通过"m_DIB.Read(file);"代码存储在 m_DIB 对象中,即 m_DIB 对象表示了位图的所有信息,对位图的各种操作即可通过 m_DIB 中的各函数实现。

第二节　位图的显示

在 CImageProcessExView 类的 OnDraw 成员函数中添加带底纹的代码,可实现位图的显示。

```
void CImageProcessExView::OnDraw(CDC * pDC)
{
    CImageProcessExDoc * pDoc = GetDocument();
    ASSERT_VALID(pDoc);
    if(pDoc->m_DIB.IsValid())
    {
        int cxDIB = (int)pDoc->m_DIB.Width();      // 位图 x 方向大小
        int cyDIB = (int)pDoc->m_DIB.Height();     // 位图 y 方向大小
        CRect rcDIB;
        rcDIB.top = rcDIB.left = 0;
        rcDIB.right = cxDIB;
        rcDIB.bottom = cyDIB;
        CRect rcDest;
        if(pDC->IsPrinting())
        {
            //获取打印页面大小
            int cxPage = pDC->GetDeviceCaps(HORZRES);
            int cyPage = pDC->GetDeviceCaps(VERTRES);
            // 获取打印机每英寸像素数
            int cxInch = pDC->GetDeviceCaps(LOGPIXELSX);
            int cyInch = pDC->GetDeviceCaps(LOGPIXELSY);


            rcDest.top = rcDest.left = 0;
            rcDest.bottom = (int)(((double)cyDIB * cxPage * cyInch)
                / ((double)cxDIB * cxInch));
            rcDest.right = cxPage;
        }
        else
```

```
        }
                rcDest = rcDIB;
        }
                pDoc->m_DIB.Paint(pDC->m_hDC, &rcDest, &rcDIB);
        }
}
```

第三节 位图的存储

类似位图读取的步骤2，在 CImageProcessExDoc 类添加虚函数 OnSaveDocument( )，并在该函数体中添加带底纹的代码，可实现位图的存储。

```
BOOL CImageProcessExDoc::OnSaveDocument(LPCTSTR lpszPathName)
{
        CFile file;
        CFileException fe;

        if (! file.Open(lpszPathName, CFile::modeCreate | CFile::modeReadWrite | CFile::shareExclusive, &fe))
        {
                ReportSaveLoadException(lpszPathName, &fe, TRUE, AFX_IDP_INVALID_FILENAME);
                return FALSE;
        }

        // 调用 SaveDIB 函数存储位图
        BOOL bSuccess = FALSE;
        TRY
        {
                BeginWaitCursor();
                bSuccess = m_DIB.Save(file);
                file.Close();
        }
        CATCH (CException, eSave)
        {
                file.Abort();
                EndWaitCursor();
                ReportSaveLoadException(lpszPathName, eSave, TRUE, AFX_IDP_FAILED_TO_SAVE_DOC);
```

```
        return FALSE;
    }
    END_CATCH

    EndWaitCursor( );
    SetModifiedFlag( FALSE );

    if ( ! bSuccess )
    {
        CString strMsg;
        strMsg.LoadString( IDS_CANNOT_SAVE_DIB );
        MessageBox( NULL, strMsg, NULL, MB_ICONINFORMATION | MB_OK );
    }
    return TRUE;
}
```

第四章 内存映射技术在大幅面图像读写的应用

文件操作是应用程序最基本的功能之一,Win32 API 和 MFC 均提供有支持文件处理的函数和类,常用的有 Win32 API 的 CreateFile( )、WriteFile( )、ReadFile( ) 和 MFC 提供的 CFile 类等。一般来说,以上这些函数可以满足大多数场合的要求,但是对于某些特殊应用领域所需要的动辄几十 GB 几百 GB 乃至几 TB 的海量存储,再以通常的文件处理方法进行处理显然是行不通的。目前,对于上述这种大文件的操作一般是以内存映射文件的方式来加以处理的。

第一节 内存映射文件技术

在 Win32 中,每个进程有自己的地址空间,一个进程不能轻易地访问另一个进程地址空间中的数据,所以不能像 16 位 Windows 那样做。内存映射文件是由一个文件到进程地址空间的映射。通过内存映射文件可以保留一个地址空间的区域,同时将物理存储器提交给此区域,只是内存文件映射的物理存储器来自一个已经存在于磁盘上的文件,如同将整个文件从磁盘加载到内存。由此可以看出,使用内存映射文件处理存储于磁盘上的文件时,将不必再对文件执行 I/O 操作,这意味着在对文件进行处理时将不必再为文件申请并分配缓存,所有的文件缓存操作均由系统直接管理,由于取消了将文件数据加载到内存、数据从内存到文件的回写以及释放内存块等步骤,因此应用内存映射文件技术处理大数据量文件的效率是非常高的。对大幅面影像读取内存映射技术比传统的 I/O 读取方式有着无可比拟的优越性。

Win32 系统允许多个进程(运行在同一计算机上)使用内存映射文件来共享数据。其他共享和传送数据的技术,诸如使用 SendMessage 或者 PostMessage,都在内部使用了内存映射文件。因此内存映射文件正是解决本地多个进程间数据共享的最有效方法。

内存映射文件并不是简单的文件 I/O 操作,实际用到了 Windows 的核心编程技术——内存管理。所以,如果想对内存映射文件有更深刻的认识,必须对 Windows 操作系统的内存管理机制有清楚的认识,内存管理的相关知识非常复杂,超出了本文的讨论范畴,在此就不再赘述,感兴趣的读者可以参阅其他相关书籍。下面给出使用内存映射文件的一般方法。

使用内存映射文件一般步骤如下:

(1)通过 CreateFile( )函数创建或打开一个文件内核对象,该对象用于标识磁盘上你想用作内存映射文件的文件。在用 CreateFile( )将文件映像在物理存储器的位置通告给操作系统后,只指定了映像文件的路径,映像的长度还没有指定。

(2)通过 CreateFileMapping()函数创建一个文件映射内核对象,告诉系统文件的大小和你打算如何访问该文件。

(3)由 MapViewOfFile()函数负责通过系统的管理而将文件映射对象的全部或部分映射到进程地址空间中。

在完成了对内存映射文件的使用后,还要通过一系列的操作完成对其清除和使用过资源的释放。这部分相对比较简单,可以先通过 UnmapViewOfFile()完成从进程的地址空间撤销文件数据的映像,然后通过 CloseHandle()关闭前面创建的文件映射对象和文件对象。

这里主要对内存映射文件中的三个主要函数进行介绍。

(1)HANDLE CreateFile()函数:创建/打开一个文件内核对象,并将其句柄返回。

在调用该函数时,需要根据是否需要数据读写和文件的共享方式来设置参数 dwDesiredAccess 和 dwShareMode,错误的参数设置将会导致相应操作的失败。

(2)HANDLE CreateFileMapping()函数:创建一个文件映射内核对象,告诉系统文件的大小以及访问这个文件的方式,运行错误时返回 NULL。

函数参数:HANDLE hFile;//由 CreateFile()函数传来的句柄

LPSECURITY_ATTRIBUTES lpFileMappingAttributes;//一般设置为 NULL

DWORD flProtect;//文件读取的方式。如设置为 PAGE_READONLY,指在映射文件映射对象时,可以读取文件中的数据;

DWORD dwMaximumSizeHigh;//文件高 32 位,一般设置为 NULL

DWORD dwMaximumSizeLow;//文件低 32 位,一般设置为 NULL

LPCTSTR lpName;//一般设置为 NULL

(3)LPVOID MapViewOfFile()函数:将文件数据映射到进程地址空间。

函数参数:HANDLE hFileMappingObject;//由 CreateFileMapping()函数返回的句柄

DWORD dwDesiredAccess;//文件读取与写入的方式,如 FILE_MAP_WRITE 允许读取以及写入文件数据

DWORD dwMaximumSizeHigh;//文件高 32 位

DWORD dwMaximumSizeLow;//文件低 32 位,这两个参数指文件映射开始位置距离头文件的偏移量

DWORD dwNumberOfBytesToMap;//映射数据量的大小

MapViewOfFile()函数允许全部或部分映射文件,在映射时,需要指定数据文件的偏移地址以及待映射的长度。其中,文件的偏移地址由 DWORD 型的参数 dwFileOffsetHigh 和 dwFileOffsetLow 组成的 64 位值来指定,而且必须是操作系统的分配粒度的整数倍,对于 Windows 操作系统,分配粒度固定为 64KB。当然,也可以通过如下代码来动态获取当前操作系统的分配粒度:

SYSTEM_INFO sinf;

GetSystemInfo(&sinf);

DWORD dwAllocationGranularity = sinf.dwAllocationGranularity;

参数 dwNumberOfBytesToMap 指定了数据文件的映射长度,这里需要特别指出的是,对于 Windows 9x 操作系统,如果 MapViewOfFile()无法找到足够大的区域来存放整个文件映

射对象,将返回空值(NULL);但是在 Windows 2000 下,MapViewOfFile( )只需要为必要的视图找到足够大的一个区域即可,而无需考虑整个文件映射对象的大小。

第二节　基于内存映射文件的文件读取示例

下面以一个统计影像均值的实例,具体介绍内存映射技术在大幅面影像读取方面的应用。实例中影像大小为 26928×31668 像素,为了方便对影像进行内存映射,先将影像转存为 Raw 格式。

```
#include "stdio.h"
#include "windows.h"
void main( )
{
    SYSTEM_INFO sinf;
    GetSystemInfo( &sinf);//获取操作系统分配粒度
    HANDLE hFile = ::CreateFile("test.raw", GENERIC_READ, FILE_SHARE_READ, NULL, OPEN_EXISTING, FILE_FLAG_SEQUENTIAL_SCAN, NULL);//影像文件名为"test.raw"
    HANDLE hFileMapping = ::CreateFileMapping(hFile, NULL, PAGE_READONLY, NULL, NULL, NULL);
    DWORD dwFileSizeHigh;
    __int64 qwFileSize = GetFileSize(hFile, &dwFileSizeHigh);
    qwFileSize += (((__int64)dwFileSizeHigh) << 32);//获取影像文件大小
    __int64 qwFileOffset = 0;//文件起始位置
    DWORD dwBytesInBlock = 1000 * sinf.dwAllocationGranularity;//映射长度
    double dMean = 0;
    while (qwFileOffset < qwFileSize)
    {
BYTE * pbFile = (BYTE *)::MapViewOfFile(hFileMapping, FILE_MAP_READ, (DWORD)(qwFileOffset>>32), (DWORD)(qwFileOffset&0xFFFFFFFF), (DWORD)dwBytesInBlock);//pbFile 数据类型与文件存储类型相同,如 BYTE 型
        for (int i = 0; i < dwBytesInBlock; i++)
        {
            dMean += *(pbFile+i);
        }//统计影像均值
        qwFileOffset += dwBytesInBlock;// 每进行一次内存文件映射后文件起始位置变更为 qwFileOffset+ dwBytesInBlock
        if (qwFileOffset+dwBytesInBlock > qwFileSize)
        {
```

```
            dwBytesInBlock = qwFileSize-qwFileOffset;
         }//防止映射出界
         ::UnmapViewOfFile(pbFile);
   }
   dMean /= (double)qwFileSize;
   CloseHandle(hFileMapping);
   CloseHandle(hFile);
}
```

与传统的 I/O 读取方式对比,内存映射技术在处理大数据量文件时表现出了良好的性能,在数据读取方面,内存映射技术比传统的 I/O 读取方式有着无可比拟的优越性。

第五章 课间实习

☞ **学生实习注意事项**

【实习要求】

(1)学生应熟悉 C 或 VC++程序设计语言,熟悉预备知识中的内容;

(2)实习前应认真复习数字图像处理相关理论知识和思考算法的设计;

(3)每人固定用一台计算机,以班级学号名建立文件夹,将自己的实习成果存放在自己的文件夹中;

(4)实习完毕,每个同学撰写一份实习报告,按时交给指导老师。

【实习考核评分方法】

学生实习成绩考核包括考勤、实习完成情况、实习报告或成果汇报等方面,各部分所占比例如下:

(1) 考勤(迟到、早退、旷课扣分)(10 分);

(2) 实习完成情况(60 分);

(3) 实习报告(课间实习 30 分);

(4) 综合实习报告(20 分),实习成果汇报(10 分)。

实习一 灰度图像直方图统计

【实习目的】

在学习灰度图像直方图的概念、计算方法、性质和相关应用基础上,学生应用 Photoshop 软件和编写灰度图像直方图统计程序,能初步掌握 Photoshop 软件操作、图像文件格式读写与图像数据处理,提高学生兴趣和编程能力,巩固所学知识。

【实习内容】

1. 利用 Photoshop 显示图像的灰度直方图,从直方图了解图像平均明暗度和对比度等信息。

2. 要求使用 C 或 C++语言编写灰度图像直方图统计的程序,并计算图像的均值和标准差。

【实习步骤】

1.使用 Photoshop 显示直方图。

(1)点击"文件"→"打开",打开一幅图像;

(2)点击"图像"→"直方图",显示图像的直方图;

（3）对图像做增强处理，例如选择"图像"→"调整"→"自动对比度"对图像进行灰度拉伸，然后再显示直方图，观察它的变化。

2.用 C 或 C++编写显示直方图的程序。

真彩色图像直方图统计的程序可以用以下伪代码表示：

```
HistogramStat( bCount[ ], gCount[ ], rCount[ ])
{
    // 赋初值,bCount,gCount,rCount 分别为蓝色、绿色、红色分量统计的结果
    bCount[0:255] = 0;
    gCount[0:255] = 0;
    rCount[0:255] = 0;
    //统计各个灰度级像素的个数
    for i=0 :nHeight              // nHeight 为图像高度
        for j=0 :nWidth           // nWidth 为图像宽度
        {   blue,green,red;       //获取第 i 行、第 j 列的蓝色,绿色和红色分量
            bCount[blue]++;
            gCount[green]++;
            rCount[red]++;
        }
}
```

实习完毕后，提交一份实习报告。

【思考题】

1. 灰度直方图可以反映出一幅图像的哪些特性？
2. 灰度直方图有何用途？编程实现一种灰度直方图应用的程序。
3. 在本次实习的基础上，试编写直方图均衡的程序。

实习二　图像增强操作

【实习目的】

在熟悉数字图像增强的基本原理和方法基础上，在理论指导下，能运用 Photoshop 软件对图像进行有针对性的增强操作，对多种图像增强方法获得的结果图像进行比较与分析，进一步熟悉和掌握 Photoshop 软件操作技能，巩固所学理论知识。

【实习内容】

应用 Photoshop 软件对图像作灰度拉伸、对比度增强、直方图均衡、图像平滑、中值滤波、边缘增强、边缘检测、伪彩色增强、假彩色合成等处理。

【实习步骤】

1. 打开一幅灰度图像。
2. 灰度拉伸。

（1）线性拉伸：在"图像 → 调整 → 色阶"中，可以通过直接设置原图像灰度值的输入

范围和所需的输出范围来简单的完成某一灰度段到另一灰度段的灰度调整映射变换。

（2）曲线拉伸：在"图像→调整→曲线"中，在弹出的"曲线"对话框中，直接用鼠标拖动改变灰度输入、输出曲线的形状，就可以完成任意线形的灰度变换。

3. 对比度增强。

对比度增强可以通过"图像→调整→亮度/对比度"来直接对原图像的亮度或对比度进行调整，观察增强处理前后图像直方图的变化。

4. 直方图均衡。

直方图均衡可调用"图像→调整→色调均化"菜单项，即可达到直方图均衡的效果。

5. 图像平滑。

（1）图像的 3×3 均匀平滑可以在"滤镜→模糊→模糊"中实现，观察处理前后图像细节和边缘的变化；也可以调用"模糊"对话框中的"高斯模糊"来观察高斯平滑处理的结果，改变半径，观察图像的变化，分析高斯平滑处理的原理。

（2）通过"滤镜→其他→自定"菜单项调出模板对话框，可以输入自定义的平滑算子或其他增强算子，改变模板的大小和缩放比例，观察处理的效果。

6. 中值滤波。

（1）先使用"滤镜→杂色→添加杂色"菜单添加噪声，再使用"滤镜→杂色→中间值"中值滤波操作，设置滤波半径，观察处理结果。

（2）采用"添加杂色"菜单项中的均匀噪声和高斯噪声给图像添加噪声，分别使用中值滤波和均匀平滑，观察这两种处理的效果，比较它们的异同。

7. 边缘增强。

（1）使用"滤镜→锐化→锐化边缘"，观察图像边缘的变化，也可以使用"USM 锐化"，"进一步锐化"等其他锐化方法。

（2）使用"滤镜→其他→自定"调出模板对话框，使用教材中讲到的 Laplace 增强算子和高通滤波算子或自定义的算子，比较它们的处理效果。

8. 边缘检测。

（1）使用"滤镜→风格化"的"查找边缘"，"等高线"，"照亮边缘"等可以提取图像的边缘，改变参数，提取图像的最佳边缘。

（2）使用"滤镜→其他→自定"，输入教材讲述的边缘检测算子，分析处理的效果，比较这些算子的特点。

9. 彩色增强。

（1）密度分割：打开一幅灰度(黑白)图像，使用"图像→模式→RGB 颜色"将图像更改成真彩色模式，再使用"图像→调整→色调分离"对话框，输入密度分割的灰度级数(1~255)，使用"图像→模式→索引颜色"将图像改成索引模式，使用"图像→模式→颜色表"对每个灰度级定义一种颜色。

（2）假彩色合成：假彩色图像合成是对一幅自然色彩图像或多光谱图像通过映射函数变换成新的三基色分量，使增强图像呈现出与原图像不同的彩色。打开一幅真彩色图像，选择"图像→调整→通道混合器"，设置对话框中的参数，观察处理后图像的变化，也可以使用"图像→调整→渐变映射"来处理。

实习完毕后，提交一份实习报告。

【思考题】

1. 通过实习，你认为中值滤波和均匀平滑在去除图像噪声上各有什么特点？试比较两种方法的异同。

2. 试比较用 Laplace、Prewitt、Sobel 和梯度算子分别对同一幅灰度图像进行边缘检测获取的边缘图像有何区别？各有什么优缺点？

3. 伪彩色增强与假彩色合成有什么区别？

第二部分 提 高 篇

第六章　GDAL 开源库及其应用

第一节　GDAL 开源库简介

GDAL(Geospatial Data Abstraction Library)是一个在 MIT/X 许可协议(MIT 许可证源自麻省理工学院 Massachusetts Institute of Technology，MIT 又称 X 条款或 X11 条款)下的开源栅格空间数据转换库，它利用抽象数据模型来表达所支持的各种文件格式，是一个操作各种栅格地理数据格式的库。这个库还同时包括了操作矢量数据的 OGR(OpenGIS Simple Feature Reference Implementation)库，一般将这两个库合称为 GDAL/OGR，或者简称为 GDAL。即 GDAL 同时具备操作栅格和矢量数据的功能。由于 MIT 协议的开放性，任何人都可以基于 GDAL 库来编写自己的软件而不需要原作者的授权，因此许多著名的 GIS 软件都是用了 GDAL/OGR 库，如 ESRI 的 ArcGIS 9.2，Google 的 Google Earth 以及开源的 GRASS GIS 系统等。

从 2007 年 3 月发布 GDAL 1.1.0 版本，到 2014 年 10 月发布 GDAL 1.11.1 版本，共发布了 45 期不同版本开源代码，现仍在不断更新完善，并陆续发布。GDAL 从最初提供对多种栅格数据的支持，包括 Arc/Info ASCII Grid(asc)，GeoTiff(tiff)，Erdas Imagine Images(img)，ASCII DEM(dem)等格式。表 6-1 是 GDAL 支持的文件格式。发展为 GDAL/OGR 用来进行栅格/矢量文件格式、数据库与 Web 服务的 C++地理空间数据访问库。最新版本的主要亮点如下：

- 新的 GDAL 驱动
- KRO:支持对 KRO KOKOR 原始格式的读写 http://gdal.org/frmt_various.html#KRO
- 新的 OGR 驱动：
  - CartoDB:读写支持 http://gdal.org/ogr/drv_cartodb.html
  - GME / Google Map Engine:读写支持 http://trac.osgeo.org/gdal/wiki/GMEDriver
  - GPKG /GeoPackage:读写支持（特指矢量特性部分）http://gdal.org/ogr/drv_geopackage.html
  - OpenFileGDB:只读支持（不依赖于外部库）http://gdal.org/ogr/drv_openfilegdb.html
  - SXF:只读支持 http://gdal.org/ogr/drv_sxf.html
  - WALK:只读支持 http://gdal.org/ogr/drv_walk.html
  - WasP.map:读写支持 http://gdal.org/ogr/drv_wasp.html
- 显著优化的驱动：GML，LIBKML

- RFC 40：增强的 RAT 支持 http://trac.osgeo.org/gdal/wiki/rfc40_enhanced_rat_support
- RFC 41：多地理字段支持 http://trac.osgeo.org/gdal/wiki/rfc41_multiple_geometry_fields
- RFC 42：OGR Layer laundered field lookup http://trac.osgeo.org/gdal/wiki/rfc42_find_laundered_fields
- RFC 43：添加函数 GDALMajorObject::GetMetadataDomainList()
- RFC 45：将 GDAL 数据库和栅格波段映射到虚拟内存 http://trac.osgeo.org/gdal/wiki/rfc45_virtualmem
- 更新到 EPSG 8.2 数据库
- 增加 OpenCL 支持，可以使用 GPU 进行处理
- 增加 CNSDTF 的格网 grd 格式和 CNSDTF 的矢量 vct 格式

表 6-1　　　　　　　　　　　GDAL 支持的文件格式

格式简称	格式全称（扩展名）	可写入	地理参考	支持文件最大数据量	默认编译
AAIGrid	Arc/Info ASCII Grid	Yes	Yes	2GB	Yes
ADRG	ADRG/ARC Digitilized Raster Graphics (.gen/.thf)	Yes	Yes	—	Yes
AIG	Arc/Info Binary Grid (.adf)	No	Yes	—	Yes
AIRSAR	AIRSAR Polarimetric	No	No	—	Yes
BLX	Magellan BLX Topo (.blx, .xlb)	Yes	Yes	—	Yes
BMP	Microsoft Windows Device Independent Bitmap (.bmp)	Yes	Yes	4GiB	Yes
BSB	BSB Nautical Chart Format (.kap)	No	Yes	—	Yes, can be disabled
BT	VTP Binary Terrain Format (.bt)	Yes	Yes	—	Yes
CEOS	CEOS	No	No	—	Yes
COASP	DRDC COASP SAR Processor Raster	No	No	—	Yes
COSAR	TerraSAR-X Complex SAR Data Product	No	No	—	Yes
CPG	Convair PolGASP data	No	Yes	—	Yes
DIMAP	Spot DIMAP (.dim)	No	Yes	—	Yes
DIPEx	ELAS DIPEx	No	Yes	—	Yes

续表

格式简称	格式全称（扩展名）	可写入	地理参考	支持文件最大数据量	默认编译
DODS	DODS / OPeNDAP	No	Yes	—	No, needs libdap
DOQ1	First Generation USGS DOQ (.doq)	No	Yes	—	Yes
DOQ2	New Labelled USGS DOQ (.doq)	No	Yes	—	Yes
DTED	Military Elevation Data (.dt0, .dt1, .dt2)	Yes	Yes	—	Yes
ECW	ERMapper Compressed Wavelets (.ecw)	Yes	Yes		No, needs ECW SDK
EHdr	ESRI .hdr Labelled	Yes	Yes	No limits	Yes
ELAS	NASA ELAS	Yes	Yes	—	Yes
ENVI	ENVI .hdr Labelled Raster	Yes	Yes	No limits	Yes
ERS	ERMapper (.ers)	Yes	Yes		Yes
ESAT	Envisat Image Product (.n1)	No	No		Yes
FAST	EOSAT FAST Format	No	Yes		Yes
FIT	FIT	Yes	No	—	Yes
FITS	FITS (.fits)	Yes	No	—	No, needs libcfitsio
FujiBAS	Fuji BAS Scanner Image	No	No		Yes
GENBIN	Generic Binary	No	No	—	Yes
GFF	GSat File Format	No	No	—	Yes
GIF	Graphics Interchange Format (.gif)	Yes	No	2GB	Yes (internal GIF library provided)
GRIB	WMO GRIB1/GRIB2 (.grb)	No	Yes	2GB	Yes, can be disabled
GMT	GMT Compatible netCDF	Yes	Yes	2GB	No, needs libnetcdf
GRASS	GRASS Rasters	No	Yes	—	No, needs libgrass
GSAG	Golden Software ASCII Grid	Yes	No	—	Yes
GSBG	Golden Software Binary Grid	Yes	No	4GiB (32767× 32767 of 4 bytes each +56 byte header)	Yes

续表

格式简称	格式全称（扩展名）	可写入	地理参考	支持文件最大数据量	默认编译
GS7BG	Golden Software Surfer 7 Binary Grid	No	No	4GiB	Yes
GSC	GSC Geogrid	Yes	No	—	Yes
GTiff	TIFF / BigTIFF / GeoTIFF（.tif）	Yes	Yes	4GiB for classical TIFF / No limits for BigTIFF	Yes (internal libtiff and libgeotiff provided)
GXF	GXF - Grid eXchange File	No	Yes	4GiB	Yes
HDF4	Hierarchical Data Format Release 4	Yes	Yes	2GiB	No, needs libdf
HDF5	Hierarchical Data Format Release 5	Yes	Yes	2GiB	No, needs libhdf5
HFA	Erdas Imagine（.img）	Yes	Yes	No limits	Yes
IDA	Image Display and Analysis	Yes	Yes	2GB	Yes
ILWIS	ILWIS Raster Map（.mpr,.mpl）	Yes	Yes	—	Yes
INGR	Intergraph Raster	Yes	Yes	2GiB	Yes
ISIS2,ISIS3	USGS Astrogeology ISIS cube	No	Yes	—	Yes
JAXAPALSAR	JAXA PALSAR Product Reader	No	No	—	Yes
JDEM	Japanese DEM（.mem）	No	Yes	—	Yes
JPEG	JPEG JFIF（.jpg）	Yes	Yes	4GiB (max dimentions 65500×65500)	Yes (internal libjpeg provided)
JPEG2000	JPEG2000（.jp2, .j2k）	Yes	Yes	2GiB	No, needs libjasper
JP2KAK	JPEG2000（.jp2, .j2k）	Yes	Yes	No limits	No, needs Kakadu library
JP2ECW	JPEG2000（.jp2, .j2k）	Yes	Yes	500MB	No, needs ECW SDK
JP2MrSID	JPEG2000（.jp2, .j2k）	Yes	Yes		No, needs MrSID SDK
L1B	NOAA Polar Orbiter Level 1b Data Set	No	Yes	—	Yes

续表

格式简称	格式全称 （扩展名）	可写入	地理参考	支持文件 最大数据量	默认编译
LAN	Erdas 7.x .LAN and .GIS	No	Yes	2GB	Yes
LCP	FARSITE v.4 LCP Format	No	Yes		Yes
Leveller	Daylon Leveller Heightfield	No	Yes	2GB	Yes
MEM	In Memory Raster	Yes	Yes	2GiB	Yes
MFF	Vexcel MFF	Yes	Yes	No limits	Yes
MFF2（HKV）	Vexcel MFF2	Yes	Yes	No limits	Yes
MrSID	Multi-resolution Seamless Image Database	No	Yes	—	No, needs MrSID SDK
MSG	Meteosat Second Generation	No	Yes		No, needs msg library
MSGN	EUMETSAT Archive native (.nat)	No	Yes		Yes
NDF	NLAPS Data Format	No	Yes	No limits	Yes
NITF	NITF	Yes	Yes	4GB	Yes
netCDF	NetCDF	Yes	Yes	2GB	No, needs libnetcdf
OGDI	OGDI Bridge	No	Yes	—	No, needs OGDI library
PAux	PCI .aux Labelled	Yes	No	No limits	Yes
PCIDSK	PCI Geomatics Database File	Yes	Yes	No limits	Yes
PCRaster	PCRaster (.map)	Yes	Yes		No, needs libcdf
PDS	NASA Planetary Data System	No	Yes	—	Yes
PGCHIP	Postgis CHIP raster	Yes	Yes	—	No, needs PostgreSQL library and Postgis headers
PNG	Portable Network Graphics (.png)	Yes	No		Yes (internal libpng provided)
PNM	Netpbm (.ppm,.pgm)	Yes	No	No limits	Yes
RIK	Swedish Grid RIK (.rik)	No	Yes	4GB	Yes

续表

格式简称	格式全称（扩展名）	可写入	地理参考	支持文件最大数据量	默认编译
RMF	Raster Matrix Format（*.rsw,.mtw）	Yes	Yes	4GB	Yes
RPFTOC	Raster Product Format/RPF（.toc）	No	Yes	—	Yes
RS2	RadarSat2 XML（.xml）	No	Yes	4GB	Yes
RST	Idrisi Raster	Yes	Yes	No limits	Yes
SAR_CEOS	SAR CEOS	No	Yes	—	Yes
SDE	ArcSDE Raster	No	Yes	—	No, needs ESRI SDE
SDTS	USGS SDTS DEM（.DDF）	No	Yes	—	Yes
SGI	SGI Image Format	Yes	Yes	—	Yes
SRTMHGT	SRTM HGT Format	Yes	Yes	—	Yes
TERRAGEN	Terragen Heightfield（.ter）	Yes	No	—	Yes
TSX	TerraSAR-X Product	Yes	No	—	Yes
USGSDEM	USGS ASCII DEM（.dem）	No	Yes	—	Yes
VRT	GDAL Virtual（.vrt）	No	Yes	—	Yes
WCS	OGC Web Coverage Server	No	Yes	—	No, needs libcurl
WMS	OGC Web Map Server	No	Yes	—	No, needs libcurl
XPM	X11 Pixmap（.xpm）	Yes	No	—	Yes

GDAL 使用抽象数据模型（Abstract Data Model）来解析它所支持的数据格式，如图 6-1 所示，抽象数据模型包括数据集（dataset）、坐标系统、仿射地理坐标转换（Affine Geo Transform）、大地控制点（GCPs）、元数据（Metadata）、栅格波段（Raster Band）、颜色表（Color Table）、子数据集域（Subdatasets Domain）、图像结构域（Image_Structure Domain）、XML 域（XML:Domains）。

GDALMajorObject 类：带有元数据的对象。

GDALDdataset 类：通常是从一个栅格文件中提取的相关联的栅格波段集合和这些波段的元数据；GDALDdataset 也负责所有栅格波段的地理坐标转换（georeferencing transform）和坐标系定义。

GDALDriver 类：文件格式驱动类，GDAL 会为每一个所支持的文件格式创建一个该类的实体,来管理该文件格式。

GDALDriverManager 类:文件格式驱动管理类,用来管理 GDALDriver 类。

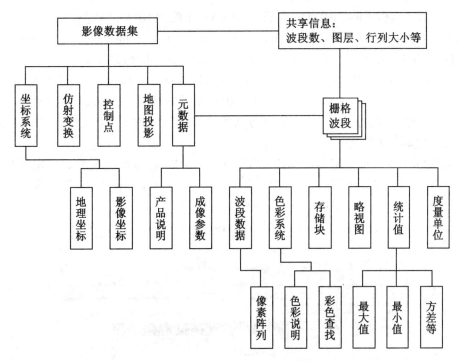

图 6-1　GDAL 开源库栅格模型

第二节　GDAL 开源库安装及其应用

一、GDAL 开源库安装

GDAL 库现有一系列版本,GDAL 库各版本在 Windows 或者其他操作系统下安装步骤大致相同。下面以 gdal1.3.0 为例,给出 GDAL 开源库在 Windows 平台下的安装步骤。

(1)下载 GDAL 的安装文件(下载网址 http://download.osgeo.org/gdal/),解压到某目录下,如 C:\gdalsrc 下。

(2)假定 VC6 安装在默认目录 C:\Program Files\Microsoft Visual Studio。打开控制台窗口,输入 cd C:\Program Files\Microsoft Visual Studio\VC98\Bin,在此目录运行 VCVARS32.BAT,然后回到 C:\gdalsrc 下。

(3)运行命令 nmake /f makefile.vc,对 GDAL 开源库进行编译。编译完成后,用记事本打开文件 nmake.opt,修改 GDAL_HOME = 这一行,确定最终 GDAL 的安装目录。比如安装在 C:\GDAL,那么这一行修改为 GDAL_HOME = "C:\GDAL"。

(4)回到目录 C:\gdalsrc 下,先执行 nmake /f makefile.vc install,再执行 nmake /f makefile.vc devinstall。这样在 C:\GDAL 目录下会生成 bin、include、lib 目录。

上述是 GDAL1.3.0 版本在 Windows 平台下成功安装的步骤。为了更简便、快速地进行 GDAL 开源库的安装,读者可以在网上搜索并下载已经编译好的 GDAL 库。

为了在 VC 编程时调用 GDAL 库,首先需在 VC 下配置 GDAL 环境。

以 VC6 为例,只需在目录 Tools\Options\Directories 中配置 Include Files,Library Files 和 Source Files 路径。或者使用固定路径的方法,首先将 include 目录复制到当前新建工程目录;将 bin 目录下的 gdal13.dll 直接放置在 Debug 目录;将 lib 目录放在新建工程目录;然后在 VC 工程下点击"Project->Setting",在弹出的对话框中点击"Link",在"Object/library modules"栏下输入"lib/gdal_i.lib"。这种方法的好处在于新建的 VC6 工程环境不会随着 PC 机环境的变化而发生变化。

下面简单介绍如何在 Visual Studio 2010 上对 GDAL 进行环境配置。首先打开 Visual Studio 2010,建立 MFC 应用程序框架,然后在"View→Property Pages"对话框中,找到 "Configuration Properties→C/C++→General",在右侧的"Additional Include Directories"中将 GDAL 的 include 文件路径填写到输入框,如图 6-2 所示。

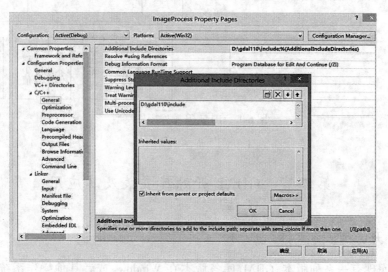

图 6-2　Visual Studio C++使用 GDAL 头文件配置

继续找到"Configuration Properties→Linker→ General",在右侧的"Additional Library Directories"中将 GDAL 的 lib 文件夹路径填写完整,如图 6-3 所示。

最后在"Configuration Properties→Linker→Input"右侧的"Additional Dependencies"中将 gdal_i.lib 填写完整。然后点击"应用","确定"按钮即可。至此,使用 GDAL 的环境全部搭建完成,剩下的就是在代码中使用 GDAL 了。

二、应用实例

在运用 GDAL 开源库之前,首先需要做两件事:一是在 VC 工程中要包括 GDAL 开源库

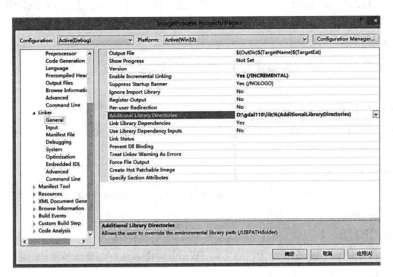

图 6-3 Visual Studio C++使用 GDAL 库文件配置

的头文件,二是添加 GDALAllRegister( )函数,实现对 GDAL 开源库所有已知驱动的注册。

1. GDAL 开源库在影像读写中的应用实例

以 IMG 格式图像为例,运用 GDAL 开源库,将 IMG 图像格式保存为 TIFF 图像格式。程序如下:

```
#include "stdio.h"
#include "gdal/gdal.h"
#include "gdal/gdal_priv.h"
#include "afx.h"
void main( )
{
    GDALAllRegister( );
    GDALDataset  * pDatasetRead;
    GDALDataset  * pDatasetSave;
    GDALDriver   * pDriver;
    int lWidth, lHeight, nBands;
    char filename[80];
    printf("Usage: main <image-file-name>\n\7");
    scanf("%s", filename);
    pDatasetRead = (GDALDataset * )GDALOpen(filename, GA_ReadOnly);
    lWidth = pDatasetRead->GetRasterXSize( );//获取图像宽度
    lHeight = pDatasetRead->GetRasterYSize( );//获取图像高度
    nBands = pDatasetRead->GetRasterCount( );//获取图像波段数
```

```cpp
    GDALDataType datatype = pDatasetRead->GetRasterBand(1)->GetRasterDataType
();//获取图像格式类型
    void *pData;
    switch(datatype)//为图像分配内存
    {
    case GDT_Byte:
        pData = (BYTE *)CPLMalloc(lWidth * lHeight * nBands * sizeof(BYTE));
        break;
    case GDT_UInt16:
        pData = (WORD *)CPLMalloc(lWidth * lHeight * nBands * sizeof(WORD));
        break;
    case GDT_Int16:
        pData = (short *)CPLMalloc(lWidth * lHeight * nBands * sizeof(short));
        break;
    case GDT_UInt32:
        pData = (DWORD *)CPLMalloc(lWidth * lHeight * nBands * sizeof(DWORD));
        break;
    case GDT_Int32:
        pData = (int *)CPLMalloc(lWidth * lHeight * nBands * sizeof(int));
        break;
    case GDT_Float32:
        pData = (float *)CPLMalloc(lWidth * lHeight * nBands * sizeof(float));
        break;
    case GDT_Float64:
        pData = (double *)CPLMalloc(lWidth * lHeight * nBands * sizeof(double));
        break;
    default:
        break;
    }
    pDatasetRead->RasterIO(GF_Read, 0, 0, lWidth, lHeight, pData, lWidth, lHeight,
datatype, nBands, NULL, 0, 0, 0);//读取图像数据
    pDriver=GetGDALDriverManager()->GetDriverByName("GTIFF");
    char **papszOptions = pDriver->GetMetadata();
    pDatasetSave=pDriver->Create("result.tif",lWidth,lHeight,nBands,datatype,
papszOptions);
    pDatasetSave->RasterIO(GF_Write, 0, 0, lWidth, lHeight, pData, lWidth,
lHeight, datatype, nBands, NULL, 0, 0, 0);//创建图像数据
```

```
        CPLFree(pData);//释放内存
        GDALClose(pDatasetRead);//关闭数据集
        GDALClose(pDatasetSave);//关闭数据集
    }
```

程序中出现的 papszOptions 表示图像格式属性,可以通过 CSLAddNameValue 为图像格式添加属性。但并不是每一种图像格式都支持 GDAL 创建,比如 PNG 格式。一般建议先建立一个过渡的 TIFF 文件,然后再用 CreateCopy 来创建需要的图像格式。注意,在最后生成新图像后,一定要关闭数据集,否则图像不会生成。

2. 利用 GDAL 开源库在 VC 6.0 环境下设计一个图像处理通用框架

图像文件格式多种多样,第一章预备知识中介绍了工程创建和 CDib 类的操作实现,但其局限性在于只能够对 BMP 位图进行处理,即只能对单色、16 色、256 色或三波段彩色图像数据进行处理。对于 TIFF、IMG 等格式图像、高比特位图像数据处理是不适用的,为此给出如何利用 GDAL 开源库在 VC 6.0 环境下设计一个图像处理通用框架,实现图像读取、显示和存储等基本功能,以飨读者。

在 GDAL 开源库的基础上设计一个图像处理通用框架,就是要创建一个类似于 CDib 类的 CImage 类。下面给出 CImage 类的具体实现步骤和代码。

(1) CImage 类的定义。

```
//image.h
#ifndef _IMAGE_H_
#define _IMAGE_H_
#include "../gdal.h"
#include "../gdal_priv.h"//GDAL 头文件的调用,具体路径依据 GDAL 环境配置
#include <cv.h>
#include <highgui.h>//添加 OpenCV 头文,为后面实习备用
class CImage
{
public:
    CImage();
    virtual ~CImage();
    void LoadDataSet(LPCTSTR lpszPathName);//导入数据并读取文件
    void SaveImage(CString lpszPathName);//保存影像
    BOOL Paint(HDC hDC, LPRECT lpDCRect, LPRECT lpDIBRect);
    int GetWidth(){return nWidth;}//获取影像宽度
    int GetHeight(){return nHeight;}//获取影像高度
    BOOL IsValid();
private:
    void CreateDIB();//创建 DIB 位图
    CString GetFileGeshi(CString str);//获取影像格式
```

```cpp
        void ReadImage();//读取影像数据
    template<class T>
        void ReadImage(T);//通过模板函数实现float等多种存储类型图像的处理
    template<class T>
        void SaveImage(T, CString lpszPathName);
public：
    GDALDataset  * pDataSet;//影像数据集
    BYTE  * pData;//保存影像数据
private：
    GDALDataType datatype;
    BITMAPINFO  * bitinfor;
    BYTE  * myDib;
    BYTE  * m_pDIBs;
    LONG lLineBYTES;
    int nWidth;
    int nHeight;
    int nBands;
    BOOL bFileFlag;
};
#endif
```

(2) CImage 类的实现。

```cpp
#include "stdafx.h"
#include "Image.h"
CImage::CImage()
{
    pDataSet = NULL;
    myDib = NULL;
    pData = NULL;
}


CImage::~CImage()
{//释放内存
    if(pDataSet ! = NULL)
    {
        GDALClose(pDataSet);
        pDataSet = NULL;
    }
    if(myDib ! = NULL)
```

```cpp
        {
            delete []myDib;
            myDib = NULL;
        }
        if(pData != NULL)
        {
            delete []pData;
            pData = NULL;
        }
}
BOOL CImage::IsValid()
{//判断影像是否成功导入
    if(pDataSet != NULL)
    {
        return TRUE;
    }
    else return FALSE;
}
CString CImage::GetFileGeshi(CString str)//获取影像存储格式
{
    if(str == "jpg" || str == "jpeg")
        return "JPEG";
    if(str == "bmp")
        return "BMP";
    if(str == "tiff" || str == "tif")
        return "GTIFF";
    if(str == "gif")
        return "GIF";
    if(str == "png")
        return "PNG";
    if(str == "img")
        return "HFA";
    return str;

}
void CImage::LoadDataSet(LPCTSTR lpszPathName)
{
    GDALAllRegister();
```

```
        pDataSet = (GDALDataset * )GDALOpen(lpszPathName,GA_ReadOnly);
        if (pDataSet == NULL)
        {
            AfxMessageBox("读取图像出错");
            return;
        }
        nWidth = pDataSet->GetRasterXSize();
        nHeight = pDataSet->GetRasterYSize();
        nBands = pDataSet->GetRasterCount();
        datatype = pDataSet->GetRasterBand(1)->GetRasterDataType();
        CreateDIB();
        ReadImage();
}
void CImage::CreateDIB()//创建 DIB 位图
{
        BITMAPINFOHEADER *hdr;
        if (nBands >= 3)
        {
            lLineBYTES = (nWidth*24+31)/32*4;
            myDib = new BYTE[sizeof(BITMAPINFOHEADER)+lLineBYTES * nHeight];
            memset(myDib, 0, sizeof(BITMAPINFOHEADER)+lLineBYTES * nHeight);
            m_pDIBs = myDib + sizeof(BITMAPINFOHEADER);
            hdr = (BITMAPINFOHEADER * )myDib;
            hdr->biBitCount = 24;
            hdr->biClrImportant = 0;
            hdr->biClrUsed = 0;
            hdr->biCompression = BI_RGB;
            hdr->biHeight = nHeight;
            hdr->biPlanes = 1;
            hdr->biSize = 40;
            hdr->biSizeImage = lLineBYTES * nHeight;
            hdr->biWidth = nWidth;
            hdr->biXPelsPerMeter = 0;
            hdr->biYPelsPerMeter = 0;
        }
        else
        {
            lLineBYTES = (nWidth*8+31)/32*4;
```

```
            myDib = new BYTE[sizeof(BITMAPINFOHEADER) + 256 * sizeof
(RGBQUAD) + lLineBYTES * nHeight];
            memset(myDib, 0, sizeof(BITMAPINFOHEADER) + 256 * sizeof
(RGBQUAD) + lLineBYTES * nHeight);
            m_pDIBs = myDib + sizeof(BITMAPINFOHEADER) + 256 * sizeof
(RGBQUAD);
            hdr = (BITMAPINFOHEADER *)myDib;
            hdr->biBitCount = 8;
            hdr->biClrImportant = 0;
            hdr->biClrUsed = 0;
            hdr->biCompression = BI_RGB;
            hdr->biHeight = nHeight;
            hdr->biPlanes = 1;
            hdr->biSize = 40;
            hdr->biSizeImage = lLineBYTES * nHeight;
            hdr->biWidth = nWidth;
            hdr->biXPelsPerMeter = 0;
            hdr->biYPelsPerMeter = 0;
            RGBQUAD * rgb = (RGBQUAD *)(myDib + sizeof(BITMAPINFOHEAD
ER));
            for(int i = 0; i < 256; i++)//给调色板赋值
            {
                rgb[i].rgbBlue   = i;
                rgb[i].rgbGreen = i;
                rgb[i].rgbRed   = i;
                rgb[i].rgbReserved = 0;
            }
        bitinfor = (BITMAPINFO *)myDib;//获取位图信息头结构
}
void CImage::ReadImage()
{
    switch(datatype)
    {
    case GDT_Byte:
        ReadImage(byte(0));
        break;
    case GDT_UInt16:
```

```
                ReadImage(WORD(0));
            break;
        case GDT_Int16:
            ReadImage(short(0));
            break;
        case GDT_UInt32:
            ReadImage(DWORD(0));
            break;
        case GDT_Int32:
            ReadImage(int(0));
            break;
        case GDT_Float32:
            ReadImage(float(0));
            break;
        case GDT_Float64:
            ReadImage(double(0));
            break;
        default:
            break;
    }
}
template<class T>
void CImage::ReadImage(T)
{
    pData = (BYTE *)malloc(nBands * nWidth * nHeight * sizeof(T));//根据不同的存储类型分配内存
    T *pDataTemp = (T *)pData;
    pDataSet->RasterIO(GF_Read, 0, 0, nWidth, nHeight, pDataTemp, nWidth, nHeight, datatype, nBands, NULL, 0, 0, 0);//读取图像数据
    double dMax, dMin;
    GDALRasterBand *pBand;
    if (nBands >= 3)//当影像波段数不小于3时
    {
        for (int k = 0; k < 3; k++)
        {
            pBand = pDataSet->GetRasterBand(k+1);
            pBand->GetStatistics(FALSE, TRUE, &dMin, &dMax, NULL, NULL);
            for (int j = 0; j < nHeight; j++)
```

```
            }
            for (int i = 0; i < nWidth; i++)
            {
                if (datatype == GDT_Byte)
                {
                    m_pDIBs[(nHeight-1-j) * lLineBYTES+i*3+(2-k)] = pDataTemp[j*nWidth+i+k*nWidth*nHeight];
                }
                Else//当影像存储类型不为 BYTE 时,对数据进行拉伸
                    m_pDIBs[(nHeight-1-j) * lLineBYTES+i*3+(2-k)] = (BYTE)(((T)pDataTemp[j*nWidth+i+k*nWidth*nHeight]-dMin) * 255/(dMax-dMin)+0.5);
            }
        }
        pBand->FlushCache();
    }
}
else
{
    pBand = pDataSet->GetRasterBand(1);
    pBand->GetStatistics(FALSE, TRUE, &dMin, &dMax, NULL, NULL);
    for (int j = 0; j < nHeight; j++)
    {
        for (int i = 0; i < nWidth; i++)
        {
            if (datatype == GDT_Byte)
            {
                m_pDIBs[(nHeight-1-j) * lLineBYTES+i]=pDataTemp[j*nWidth+i];
            }
            else
                m_pDIBs[(nHeight-1-j) * lLineBYTES+i] = (BYTE)(((T)pDataTemp[j*nWidth+i]-dMin) * 255/(dMax-dMin)+0.5);
        }
    }
    pBand->FlushCache();
}
}
```

```cpp
void CImage::SaveImage(CString lpszPathName)
{
    switch (datatype)
    {
    case GDT_Byte:
        SaveImage(byte(0), lpszPathName);
        break;
    case GDT_UInt16:
        SaveImage(WORD(0), lpszPathName);
        break;
    case GDT_Int16:
        SaveImage(short(0), lpszPathName);
        break;
    case GDT_UInt32:
        SaveImage(DWORD(0), lpszPathName);
        break;
    case GDT_Int32:
        SaveImage(int(0), lpszPathName);
        break;
    case GDT_Float32:
        SaveImage(float(0), lpszPathName);
        break;
    case GDT_Float64:
        SaveImage(double(0), lpszPathName);
        break;
    default:
        break;
    }
}

template<class T>
void CImage::SaveImage(T, CString lpszPathName)
{
    int index;
    index = lpszPathName.Find('.');
    CString str = lpszPathName.Right(lpszPathName.GetLength()-index-1);
    str.MakeLower();
    str = GetFileGeshi(str);
    GDALDataset * pDatasetSave;
```

GDALDriver　　＊pDriver；
　　pDriver＝GetGDALDriverManager（）->GetDriverByName（str）；
　　char ＊＊papszOptions＝pDriver->GetMetadata（）；
　　pDatasetSave ＝ pDriver － ＞ Create（lpszPathName, nWidth, nHeight, nBands, datatype, papszOptions）；
　　T ＊ pDataTemp ＝（T＊）pData；
　　pDatasetSave->RasterIO（GF_Write, 0, 0, nWidth, nHeight, pDataTemp, nWidth, nHeight, datatype, nBands, NULL, 0, 0, 0）；
　　GDALClose（pDatasetSave）；
｝
BOOL CImage::Paint（HDC hDC, LPRECT lpDCRect, LPRECT lpDIBRect）
｛
　　::SetStretchBltMode（hDC, COLORONCOLOR）；
　　::StretchDIBits（hDC,lpDCRect->left,lpDCRect->top,lpDCRect->right-lpDCRect->left,lpDCRect->bottom-lpDCRect->top,
　　　　lpDIBRect->left,lpDIBRect->top,
　　　　lpDIBRect->right-lpDIBRect->left, lpDIBRect->bottom-lpDIBRect->top, m_pDIBs, bitinfor,DIB_RGB_COLORS, SRCCOPY）；
　　return TRUE；
｝

下面是基于 GDAL 设计的图像读写、显示和存储模块及其代码。

3. 图像读取功能实现

（1）新建一个多视图文档工程,取名为 Imageprocess。在 MFC AppWizard 属性选项中将 Base View 改为 CScrollView。

（2）在 CImageprocessDoc 中定义 CImage 类型的对象。在 ClassView 视窗中的 CImageprocessDoc 类上单击右键,选择"Add Member Variable…"菜单项,在弹出的对话框中输入变量类型 CImage,变量名为 m_img 的 public 型变量,单击 OK 按钮后返回。这时你会发现系统自动在 ImageprocessDoc.h 文件中加入了下面两行代码。

　　#include "Image.h"// Added by ClassView
　　CImage m_img；

（3）在 ClassView 的 CImageprocessDoc 类上点击鼠标右键,在弹出菜单上选择"Add Virual Function",选取 OnOpenDocument,再单击 Add and Edit 按钮,系统将自动回到添加的 OnOpenDocument（）函数的定义处。在 OnOpenDocument（）函数里添加如下代码：

　　m_img.LoadDataSet（lpszPathName）；
　　if（！m_img.IsValid（））
　　｛
　　　　AfxMessageBox（"图像导入失败"）；

```
        return FALSE;
}
SetPathName(lpszPathName);
SetModifiedFlag(FALSE);
```

通过添加上述代码,当执行程序"文件/打开"菜单项时,影像信息就存储在 pDataSet 数据集上。运用 GDAL 库函数可以得到数据集的详细信息,同时影像灰度值保存于 pData。由于 pData 定义的是 BYTE 类型指针,读者在对图像进行处理时,需要将数据进行类似于 CImage 类中 SaveImage()函数里的类型转换。

4. 图像显示功能实现

在 CImageprocessView 类的 OnDraw 成员函数中添加如下代码,可实现图像的显示:

```
if(pDoc->m_img.IsValid())
{
    int cxDIB = (int)pDoc->m_img.GetWidth();    //size of dib-x
    int cyDIB = (int)pDoc->m_img.GetHeight();   //size of dib-y
    CRect rcDIB;
    rcDIB.top = rcDIB.left = 0;
    rcDIB.right = cxDIB;
    rcDIB.bottom = cyDIB;
    CRect rcDest;
    if(pDC->IsPrinting())
    {
        int cxPage = pDC->GetDeviceCaps(HORZRES);
        int cyPage = pDC->GetDeviceCaps(VERTRES);
        int cxInch = pDC->GetDeviceCaps(LOGPIXELSX);
        int cyInch = pDC->GetDeviceCaps(LOGPIXELSY);
        rcDest.top = rcDest.left = 0;
rcDest.bottom = (int)(((double)cyDIB * cxPage * cyInch)/((double)cxDIB * cxInch));
        rcDest.right = cxPage;
    }
    else    //not printer DC
    {
        rcDest = rcDIB;
    }
    pDoc->m_img.Paint(pDC->m_hDC,&rcDest,&rcDIB);
    CSize size(pDoc->m_img.GetWidth(),pDoc->m_img.GetHeight());
    SetScrollSizes(MM_TEXT,size);//更新滚动条信息
}
```

5. 图像存储功能实现

由于图像数据存储类型的不一致,如 BMP 图像只能存储 BYTE 型数据,而 TIFF、IMG 等则可以存储 float 等多种数据格式,为了防止影像存储时发生错误,可将影像存储为 TIFF、IMG 等通用格式。当然读者也可以自行设计程序,在存储图像前进行数据类型转换,使之能够存储为 BMP 格式。

在 VC 菜单 View 下点击"MFC ClassWizard"项,在 CImageprocessDoc 类中的 ID_FILE_SAVE_AS 添加函数,单击确定返回。在 OnFileSaveAs( )函数中加入如下代码:

```
CFileDialog MySaveFileDlg ( FALSE, "", NULL, OFN _ HIDEREADONLY | OFN _ OVERWRITEPROMPT, "TIFF 文件( * .tif) | * .tif || ", NULL);
    if ( MySaveFileDlg.DoModal( ) = = IDOK )
    {
        m_img.SaveImage( MySaveFileDlg.GetPathName( ));
    }
```

通过添加上述代码,执行程序"文件/另存为"菜单项就会弹出对话框,输入文件名保存,就可以实现图像的存储。

在综合实习阶段,要求实习学生在上述通用平台下设计算法。

第七章 OpenCV 开放源代码简介及其应用

第一节 OpenCV 开放源代码简介

OpenCV 这一名称包含了 Open 和 Computer Vision 二者的意思。实际上,Open 指 Open Source(开源,即开放源代码),Computer Vision 则指计算机视觉。OpenCV 是 Intel 公司的开源计算机视觉库。它由一系列 C 函数和少量 C++类构成,实现了图像处理和计算机视觉方面的很多通用算法。

OpenCV 作为一个外部函数库,可以通过在 WINDOWS、LINUX 环境下的 IDE 下配置连接,在编程开发时直接调用库中各种算法对应的函数来加快开发速度。OpenCV 尤其针对 INTEL 处理器做了优化,在 INTEL 平台下能表现出更快的速度。

OpenCV 库中的函数主要分为以下五类:

Highgui:高层图形用户接口,主要功能负责图形界面的创建、操作和销毁等。另外,图像、视频的读写等基本操作也包含在此库里。

Cxcore:视觉库进行一切处理的核心操作函数,包括基本及高等数学运算函数,点、面、矩阵等基础结构的定义,底层内存的分配、清空等存储操作函数,动态数据结构各种操作,如序列结构数据、集合结构数据、图型和树型结构数据的基本操作函数。还包括文件读写存储、错误的发现处理和与系统运行有关的函数。

Cv:图像、视频的处理函数,包括基本的几何变换、灰度与色彩处理、频域滤波、边缘轮廓提取、图像分割、金字塔采样、模式识别等,还可以通过其中的函数来对图像图形进行某种要求的解求和分析运算。还包括针对视频帧的各种处理算法:统计计算、运动分析、目标跟踪及光流场的分析、预测器的运算函数。摄像机定标和三维重建用到的基本算法也被编成函数加入其中,可以通过直接调用实现此功能。

Cvcam:有关摄像机操作的函数。

Cvaux:实验性的算法,三维跟踪,主成分分析,马尔柯夫模型等相关算法。

其中,前三个库是最常用的,后两个库较高级、抽象,是随着版本的更新增加的,也会随着计算机视觉的发展逐渐丰富起来。

在编程应用中,利用到某个功能的函数可以通过其大致类别判断在哪个包里,然后再在发布的帮助手册中寻找,帮助手册中有每个函数的详细的介绍。

OpenCV 作为开放的数字图像处理和计算机视觉软件平台,具有以下特点:

(1)开放 C 源码。

(2)基于 Intel 处理器指令集开发的优化代码。

（3）统一的结构和功能定义。

（4）强大的图像和矩阵运算能力。

（5）方便灵活的用户接口。

（6）同时支持 MSWindows 和 Linux 平台。

OpenCV 作为开放源代码的图像库，逐步成为一个通用的基础研究和产品开发平台其研究和应用，越来越引起人们的关注，也会随着研究队伍的扩大功能越来越强大，越来越完善。OpenCV 的发展对软件的开发具有重要影响。源代码是由软件命令电脑执行指定动作的程序语句，是一个软件的核心所在。开放源代码软件能风靡世界，首先是其开源的免费特性；此外，由于有全球众多编程者的参与，开源软件一般具有简约精练、资源占用少、功能集中和安全性好的特点。

计算机视觉（Computer Vision）是在数字图像处理的基础上发展起来的新兴学科，它从信息处理的层次研究视觉信息的认知过程，研究视觉信息处理的计算理论和表达与计算方法，包括图像特征提取，摄像机定标，立体视觉，运动视觉（或称序列图像分析），由图像灰度恢复三维物体形状的方法，物体建模与识别方法以及距离图像分析方法等方面。作为一门综合性的交叉学科，计算机视觉处理的领域涉及计算机科学与工程、信号处理、物理学、应用数学与统计学以及神经生理学与认知科学等方面，并在制造业、检验、文档分析、医疗诊断和军事等领域的各种智能/自主系统中有着广泛的应用。

开放源代码不仅使软件比较便宜，它更是一种转变，一种建造当代软件公司发展模式的运动。

从长远来讲，开放源代码或许是一种效率更高、效果更好的软件商业模式。因为开源作为今后软件的发展模式，必将给现有的高技术企业带来巨大的潜在利益。目前，以 Linux 为代表的开放源代码软件得到了 HP、IBM、Sun、Intel 和 Novell 等世界上几乎所有大计算机软、硬件厂商的重视和支持，他们纷纷启动了开发和使用开放源代码软件的项目。因为开源已经日益成为一个不可阻挡的世界潮流，无论从软件开发本身还是技术趋势来看，开放的、有众多志愿者参与的开源项目，都会大大促进产品和技术的发展，并带来巨大的潜在商业价值。正因为如此，OpenCV 在 Intel 公司的鼎力资助下，才得到长足和有保障的发展：从 1.x 版本使用 C 的 API，到 2.x 版本同时提供 C++的 API，再到最新版本 3.0 alpha，添加更多功能，速度更快，稳定性更强。

在此简单介绍 OpenCV，主要是希望能够对 OpenCV 在国内的推广有所裨益，为那些对 OpenCV 感兴趣以及刚刚开始接触 OpenCV 的人消除陌生感。

第二节　OpenCV 库的安装、配置与应用实例

下面分别介绍 OpenCV 库在 VC 和 VS2010 环境的安装、配置，以及 OpenCV 库的应用实例。

一、VC 环境下 OpenCV 库的安装、配置

首先到 http://www.opencv.org.cn 下载 OpenCV 安装程序，建议 VC6 用户下载 Open

CV1.0版本安装。下面介绍在VC6环境下的安装步骤。

若将OpenCV安装到C:\Program Files\OpenCV目录下,首先在安装时勾选"将\OpenCV\bin加入系统变量"(Add\OpenCV\bin to the systerm PATH);然后检查C:\Program Files\OpenCV\bin是否已被加入到环境变量PATH,如果没有,请加入。加入后需要注销当前Windows用户(或重启)后重新登录才生效。

在VC环境下对OpenCV库需进行配置。配置方法之一是在菜单"Tools→Opyions→Direcrtories"下的Include Files,Library Files和Source Files设置头文件路径、lib路径和src路径。配置方法之二是使用固定路径的方法。首先将OpenCV头文件和lib目录复制到当前新建工程目录下,然后将bin目录下的OpenCV动态链接库直接放置在新建工程Debug目录下。后一种方法的好处在于新建的VC6工程环境不会随着PC机环境的变化而发生变化。

二、在Visual Studio 2010环境下对OpenCV库进行配置

(1)配置OpenCV环境变量。找到计算机→(右键)属性→高级系统设置→高级(标签)→环境变量→系统里面path→编辑→在"变量值"里面添加"C:\OpenCV\bin"(里面的opencv记得换成自己的opencv路径。),如图7-1所示。

图7-1 OpenCV环境变量设置

(2)打开Visual Studio 2010,建立MFC应用程序框架,然后在View→Property Pages对话框中,找到Configuration Properties→VC++ Directories,在右侧的"Include Directories"和"Library Directories"中将OpenCV的include和lib文件路径填写到输入框,如图7-2所示。

第七章 OpenCV 开放源代码简介及其应用

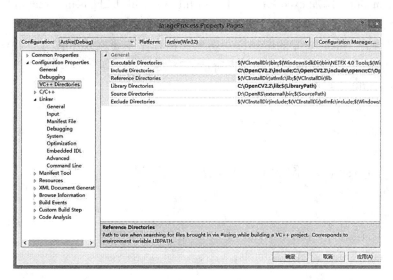

图 7-2 include 和 lib 文件配置

（3）配置连接器，在 Configuration Properties→Linker→Input，在右侧的"Additional Dependencies"中将需要的 lib 文件填入，如图 7-3 所示。

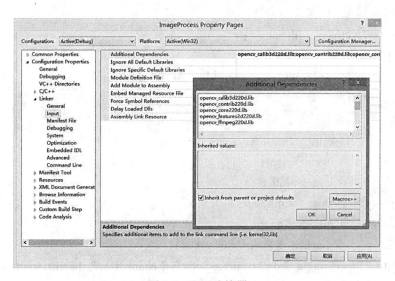

图 7-3 配置连接器

（4）点击"应用"、"确定"按钮即可。至此，使用 OpenCV 的环境全部搭建完成，剩下的就是在代码中使用 OpenCV 函数了。

三、应用实例

当创建一个基于 OpenCV 库的新工程时，只需要在 Project→Setting→link 栏下的

91

Object/library modules 添加 cxcore.lib cv.lib ml.lib cvaux.lib highgui.lib cvcam.lib,其中 lib 文件可以视自己工程需要添加。

下面给出基于 OpenCV 库的影像读写实例。该程序主要实现了读入一幅图像,对它进行反色处理,然后显示的功能。代码如下所示:

```
#include <stdlib.h>
#include <stdio.h>
#include <math.h>
#include <cv.h>
#include <highgui.h>//添加 OpenCV 头文件
int main( )
{
    IplImage * img = 0;
    int height,width,step,channels;
    uchar * data;
    int i,j,k;
    char filename[80];
    printf("Usage:main <image-file-name>\n\7");
    scanf("%s", filename);
    //导入图像
    img=cvLoadImage(filename);
    if(! img){
        printf("Could not load image file:%s\n",filename);
        exit(0);
    }
    // 获取图像数据
    height    = img->height;
    width     = img->width;
    step      = img->widthStep;
    channels  = img->nChannels;
    data      = (uchar *)img->imageData;
    printf("Processing a %dx%d image with %d channels\n",height,width,channels);
    // 创建一个窗口
    cvNamedWindow("mainWin", CV_WINDOW_AUTOSIZE);
    cvMoveWindow("mainWin", 100, 100);
    //将图像进行反色处理
    // 相当于 cvNot(img);
    for(i=0;i<height;i++) for(j=0;j<width;j++) for(k=0;k<channels;k++)
        data[i*step+j*channels+k]=255-data[i*step+j*channels+k];
```

//显示结果图像
cvShowImage("mainWin", img);
// wait for a key
cvWaitKey(0);
// 释放图像
cvReleaseImage(&img);
return 0;
}

第八章 大幅面图像分块处理

第一节 分块处理方法

对于一般的图像处理,因其数据量较小,可以一次性将图像数据读入内存进行处理;对大幅面图像而言,会遇到因计算机内存有限无法将所有数据一次性读入内存的问题,在前一部分第四章介绍了内存映射技术在大幅面图像读写的应用,可作为解决办法之一。图像数据分块处理是解决这个问题的又一方法。简单分块处理方法是一种顺序的处理过程:读入一块图像数据进行处理,然后输出结果;按类似方式处理其他块数据。由于计算机对数据的读入和写出是按磁道分块进行,简单分块处理方法增加了读写时间,还有协调分块和结果合并的问题,不能充分利用 CPU 资源,因而处理效率较低。

进程是指应用程序的一个执行程序,线程是进程内的一个执行路线,当多线程程序执行时,对应的进程包含并发执行的多个线程。利用多线程可以执行实时性、交互性很强的运算和操作,能更有效地利用系统资源,提高大幅面图像的处理效率,主要依据有以下两点:

(1)在支持多任务处理的操作系统中,任何应用程序都不像在单任务操作系统中那样独享所有系统资源,操作系统将 CPU 时间片以进程为基本单位进行分配,进程进一步划分为线程。在同一优先级下,单线程程序所分配到的 CPU 时间要比多线程程序少,而采用多线程程序能得到更多 CPU 时间。

(2)大幅面图像的处理主要包括图像运算和磁盘读/写访问两个部分。因为一次只能读取一部分数据进行处理,所以磁盘访问次数比较频繁,而访问磁盘设备速度相对较慢。若采用顺序处理方法,则只有等待磁盘访问完成后进行下一步处理,因此存在等待的过程中未充分利用 CPU 资源的问题。而采用多线程技术,则有可能使处理数据与磁盘操作两部分同时进行,既提高对 CPU 的利用率又提高数据处理效率。

多线程图像处理技术的核心思想是将待处理的图像数据逻辑地划分为多个部分,每一个部分由一单独的线程来进行处理。各线程分别读取属于自己处理范围内的数据、分别进行处理并写入对应的数据区域。各线程之间是相互独立、互不影响的。多线程图像处理中每一线程都需要对输入和输出文件进行访问,因此需要采用某种共享的策略来实现共享。常用的方法是利用 Windows 提供的同步机制,协调各线程对文件资源的访问达到共享的目的。也可通过指定文件共享访问标志来达到共享的目的。以 Win32 API 的 CreateFile 为例,如果在共享模式中指定 FIIE_SHARE_READ 标志,后续线程可使用 GENERIC_READ 标志打开该文件进行读操作;如果指定 FIIE_SHARE_WRITE 标志,后续线程可以使用 GENERIC_WRITE 标志打开该文件进行写操作。在多线程处理中,对于输入文件采用读共享,对于输出文件采用写共享,这样可以保证每个线程都能对输入输出文件进行相应的访问。这种方法不需进行线程同步,因而大大减少了额外开销,提高了处理效率。对图像数据的逻辑划分可以按高度或宽度进行,也可以对高度宽度同时进行划分。多线程处理的示意图如图 8-1 所示。

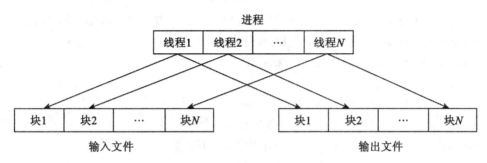

图 8-1　多线程处理示意图

第二节　分块处理算法实例

下面我们分别给出采用简单分块处理和多线程图像处理的对比实例,程序的主要功能是读入影像,延迟一定时间后再写入影像。为了方便进行操作,我们选用 8 位 BMP 图像作为处理对象。

一、简单分块处理实例代码

```
#include "stdio.h"
#include "afx.h"
BITMAPFILEHEADER bmh;
BITMAPINFOHEADER bih;
RGBQUAD RGB[256];
    void main()
    {
    char filename_in[80];
    printf("Usage: main <image-file-name>\n\7");
    scanf("%s", filename_in);
    //获取影像宽度高度信息
    HANDLE hFile_in = CreateFile(filename_in, GENERIC_READ, FILE_SHARE_READ,
NULL, OPEN_EXISTING, FILE_ATTRIBUTE_NORMAL, NULL);
    CString filename_out;
    filename_out.Format("%s", "result.bmp");
    HANDLE hFile_out = CreateFile(filename_out, GENERIC_WRITE, FILE_SHARE_
```

```cpp
WRITE, NULL, CREATE_ALWAYS, FILE_ATTRIBUTE_NORMAL, NULL);
    DWORD readsize;
    ReadFile(hFile_in, &bmh, sizeof(BITMAPFILEHEADER), &readsize, NULL);
    ReadFile(hFile_in, &bih, sizeof(BITMAPINFOHEADER), &readsize, NULL);
    ReadFile(hFile_in, RGB, 256 * sizeof(RGBQUAD), &readsize, NULL);
    WriteFile(hFile_out, &bmh, sizeof(BITMAPFILEHEADER), &readsize, NULL);
    WriteFile(hFile_out, &bih, sizeof(BITMAPINFOHEADER), &readsize, NULL);
    WriteFile(hFile_out, RGB, 256 * sizeof(RGBQUAD), &readsize, NULL);//写入文件头信息
    int lWidth = bih.biWidth;//获取图像宽度
    int lHeight = bih.biHeight;//获取图像高度
    SetFilePointer(hFile_in, bmh.bfOffBits, NULL, FILE_BEGIN);
    SetFilePointer(hFile_out, bmh.bfOffBits, NULL, FILE_BEGIN);
    for (int j = 0; j < lHeight; j++)
    {
        BYTE * pData = new BYTE[lWidth];
        DWORD readsize;
        ReadFile(hFile_in, pData, lWidth, &readsize, NULL);
        for (int i = 0; i < lWidth; i++)
        {
            if (pData[i] > 50)
            {
                pData[i] = 255;
            }
            else
                pData[i] = 0;
        }
        WriteFile(hFile_out, pData, lWidth, &readsize, NULL);
        delete []pData;
    }
    CloseHandle(hFile_in);
    CloseHandle(hFile_out);
}
```

二、多线程图像处理实例

多线程图像处理中主线程负责对图像进行分块计算,创建子线程,并将相应参数传递给每个子线程。主线程处理流程为:

(1) 读入图像的高度和宽度等基本信息。

(2) 根据指定的线程数目创建子线程列表,计算各线程处理范围并启动子线程。

(3) 查询子线程状态标志,如果所有子线程已完成处理则释放资源并退出,否则继续等待。

子线程实现具体的图像处理操作,其流程为:

(1) 打开输入和输出文件。

(2) 从输入文件读取处理范围内的部分数据,处理并写入输入文件对应位置。

(3) 数据处理完毕后,释放资源,标记子线程状态为完成状态并退出,否则转入(2)。

参数传递是主线程与子线程通信的主要方式,由于主线程启动线程时只能传递一个参数给子线程,因而多个参数的传递常通过传递结构体的地址来实现多参数传递。结构体需包含的几个区域:

(1) 子线程的状态。用于标记子线程是否已经处理完毕,首先由主线程赋为假值,子线程处理完后将其置为真值。

(2) 输入/输出文件名。如果图像运算涉及多输入/多输出,可将该区域设为字符串数组。

(3) 子线程处理区域。

具体实现代码如下所示:

```
#include "stdio.h"
#include "math.h"
#include "afx.h"
#include "process.h"
#define nThreadCount 4//线程数目
typedef struct
{
    CString filename_in;//输入文件名
    CString filename_out;//输出文件名
    int nStartY;
    int nEndY;//分块区域范围
    bool bFlag;//子线程状态标志
}threadInfo;
BITMAPFILEHEADER bmh;
BITMAPINFOHEADER bih;
RGBQUAD RGB[256];
void ThreadFunc(LPVOID lpParam)
{
    threadInfo * pInfo = (threadInfo *)lpParam;
    HANDLE hFile_in = CreateFile(pInfo->filename_in, GENERIC_READ, FILE_SHARE_READ, NULL, OPEN_EXISTING, FILE_ATTRIBUTE_NORMAL, NULL);
    HANDLE hFile_out = CreateFile(pInfo->filename_out, GENERIC_WRITE, FILE_
```

```
    SHARE_WRITE, NULL, OPEN_ALWAYS, FILE_ATTRIBUTE_NORMAL, NULL);
        int lWidth = bih.biWidth;//获取图像宽度
        SetFilePointer(hFile_in, bmh.bfOffBits+pInfo->nStartY * lWidth, NULL, FILE_
    BEGIN);
        SetFilePointer(hFile_out, bmh.bfOffBits+pInfo->nStartY * lWidth, NULL, FILE_
    BEGIN);
        for (int j = pInfo->nStartY; j < pInfo->nEndY; j++)
        {
            BYTE * pData = new BYTE[lWidth];
            DWORD readsize;
            ReadFile(hFile_in, pData, lWidth, &readsize, NULL);
            for (int i = 0; i < lWidth; i++)
            {
                if (pData[i] > 50)
                {
                    pData[i] = 255;
                }
                else
                    pData[i] = 0;
            }
            WriteFile(hFile_out, pData, lWidth, &readsize, NULL);
            delete []pData;
        }
        CloseHandle(hFile_in);
        CloseHandle(hFile_out);
        pInfo->bFlag = true;
    }
    void main()
    {
        char filename_in[80];
        printf("Usage: main <image-file-name>\n\7");
        scanf("%s", filename_in);
        CString filename_out;
        filename_out.Format("%s","result.bmp");
        HANDLE hFile_in = CreateFile(filename_in, GENERIC_READ, FILE_SHARE_
    READ, NULL, OPEN_EXISTING, FILE_ATTRIBUTE_NORMAL, NULL);
        HANDLE hFile_out = CreateFile(filename_out, GENERIC_WRITE, FILE_SHARE_
    WRITE, NULL, CREATE_ALWAYS, FILE_ATTRIBUTE_NORMAL, NULL);
```

```
    DWORD readsize;
    ReadFile(hFile_in, &bmh, sizeof(BITMAPFILEHEADER), &readsize, NULL);
    ReadFile(hFile_in, &bih, sizeof(BITMAPINFOHEADER), &readsize, NULL);
    ReadFile(hFile_in, RGB, 256 * sizeof(RGBQUAD), &readsize, NULL);
    WriteFile(hFile_out, &bmh, sizeof(BITMAPFILEHEADER), &readsize, NULL);
    WriteFile(hFile_out, &bih, sizeof(BITMAPINFOHEADER), &readsize, NULL);
    WriteFile(hFile_out, RGB, 256 * sizeof(RGBQUAD), &readsize, NULL);//写入文件头信息
    int lWidth = bih.biWidth;//获取图像宽度
    int lHeight = bih.biHeight;//获取图像高度
    //多线程图像处理
    threadInfo info[nThreadCount];
    for (int i = 0; i < nThreadCount; i++)
    {
        info[i].filename_in = filename_in;
        info[i].filename_out = filename_out;
        info[i].bFlag = false;
        info[i].nStartY = floor((float)(i * lHeight)/nThreadCount);
        info[i].nEndY = floor((float)((i+1) * lHeight)/nThreadCount);
        _beginthread(ThreadFunc, 0, &info[i]);//创建线程
    }
    for (i = 0; i < nThreadCount; i++)//判断子线程是否结束
    {
        while (! info[i].bFlag)
        {
            Sleep(500);
        }
    }
    CloseHandle(hFile_in);
    CloseHandle(hFile_out);
}
```

随着集成电路制造工艺水平的迅猛提高和计算机微体系结构设计技术的发展和创新，微处理器的发展已经迈入线程级并行的时代。线程级并行将会是下一代高性能处理器的核心技术，多线程体系结构有望在未来10年中占主体地位。

为此，研究人员曾经提出过两种不同于传统的单核处理器的结构：单片多处理器(Chip Multiprocessor, CMP)与同时多线程处理器(Simultaneous Multithreading, SMT)，这两种处理器结构可以充分利用这些应用的指令级并行性和线程级并行性，从而显著提高了应用性。

CMP是指由在单个芯片上的多个处理器核所构成的多处理器系统。CMP允许线程在

多个处理器核上并行执行,从而利用线程级并行性来提高系统性能。CMP 的优势主要是处理器核可以很简单,易于获得较高的主频,同时缩短了设计和验证时间。CMP 存在的主要问题是由于单片多处理器系统的资源是采用划分方式的,当没有足够的线程时,资源就浪费了。

SMT 结构的基本思想是:在一个时钟周期内发射多个线程的指令到功能部件上执行,SMT 的优势是结合了超标量和多线程处理器的特点。SMT 允许在一个时钟周期内执行来自不同线程的多条指令,因此在一个时钟周期内,SMT 能够同时利用程序的线程级并行和指令级并行来消除水平浪费。SMT 允许多个活动线程的组合来发射指令,当由于长延迟操作或者资源冲突只有一个活动线程时,该线程能够使用所有可获得的发射槽来消除垂直浪费。SMT 存在的主要问题是使得指令发射阶段变得比较复杂;随着指令发射宽度的增加,模块和电路的延迟越来越大,不易获得较高的主频;另外,多个线程共享同一个一级 cache、TLB 和分支预测逻辑,将会导致冲突,使得 cache 不命中和分支预测错误率增加,限制了整个处理器性能的提升。

多核多线程处理器是通过支持单片多处理器 CMP 和同时多线程 SMT 的组合来实现的。多核多线程处理器由多个简单的同时多线程处理器核构成,它提供了一种更加简单有效的方法去提高集成度。它不同于超标量处理器通过硬件来提取指令级的并行,是通过编译器的支持,多核多线程处理器可以提供一种线程级的并行。由于它由多个简单的同时多线程处理器,所以它就可以拥有单片多处理器主频高、设计和验证时间短的优势,又拥有同时多线程资源利用率高的优势,从而大大提高程序的运行效率。目前,越来越多的芯片生产商和研究机构都将注意力放在了多核多线程处理器的研究上。

多核多线程在通用处理器设计方面已经开始引领潮流了。在不久的将来,多核多线程处理器将占领整个计算机领域,从笔记本电脑和台式电脑再到服务器和超级计算机。

第九章 图像并行处理技术

第一节 OpenMP 简介与应用实例

一、OpenMP 简介

OpenMP 是由 OpenMP Architecture Review Board 牵头提出的,是一个支持共享存储并行设计的库,特别适宜多核 CPU 上的并行程序设计。OpenMP 支持的编程语言包括 C 语言、C++和 Fortran;而支持 OpenMP 的编译器包括 Sun Compiler、GNU Compiler 和 Intel Compiler 等。OpenMP 提供了对并行算法的高层的抽象描述,程序员通过在源代码中加入专用的 pragma 来指明自己的意图,由此编译器可以自动将程序进行并行化,并在必要之处加入同步互斥以及通信。当选择忽略这些 pragma,或者编译器不支持 OpenMP 时,程序又可退化为通常的程序(一般为串行),代码仍然可以正常运作,只是不能利用多线程来加速程序执行。

OpenMP 提供的这种对于并行描述的高层抽象降低了并行编程的难度和复杂度,这样程序员可以把更多的精力投入到并行算法本身,而非其具体实现细节。对基于数据分集的多线程程序设计,OpenMP 是一个很好的选择。同时,使用 OpenMP 也提供了更强的灵活性,可以较容易地适应不同的并行系统配置。线程粒度和负载平衡等是传统多线程程序设计中的难题,但在 OpenMP 中,OpenMP 库从程序员手中接管了这两方面的部分工作。

但是,作为高层抽象,OpenMP 并不适合需要复杂的线程间同步和互斥的场合。OpenMP 的另一个缺点是不能在非共享内存系统(如计算机集群)上使用。在这样的系统上,MPI 使用较多。

二、OpenMP 应用实例

在 VC8.0 中项目的属性对话框中,左边框里的"配置属性"下的"C/C++ -> Language"选项卡中将"OpenMP Support"项设置为"Yes",然后在需要使用 OpenMP 函数的 cpp 文件中引用#include <omp.h>,这样设置就完成了。注意,Debug 模式和 Release 模式都需要设置 OpenMP Support 选项。

先看一个简单的使用了 OpenMP 的程序:
```
int main( int argc, char * argv[ ] )
{
#pragma omp parallel for
    for ( int i = 0; i < 10; i++ )
```

parallel sections,parallel 和 sections 两个语句的结合。

critical,用在一段代码临界区之前。

single,用在一段只被单个线程执行的代码段之前,表示后面的代码段将被单线程执行。

flush,用来保证线程的内存临时视图和实际内存保持一致,即各个线程看到的共享变量是一致的。

barrier,用于并行区内代码的线程同步,所有线程执行到 barrier 时要停止,直到所有线程都执行到 barrier 时才继续往下执行。

atomic,用于指定一块内存区域被制动更新。

master,用于指定一段代码块由主线程执行。

ordered,用于指定并行区域的循环按顺序执行。

threadprivate,用于指定一个变量是线程私有的。

OpenMP 除上述指令外,还有一些库函数,下面列出几个常用的库函数:

  omp_get_num_procs,返回运行本线程的多处理机的处理器个数。

  omp_get_num_threads,返回当前并行区域中的活动线程个数。

  omp_get_thread_num,返回线程号。

  omp_set_num_threads,设置并行执行代码时的线程个数。

omp_init_lock,初始化一个简单锁。

omp_set_lock,上锁操作。

omp_unset_lock,解锁操作,要和 omp_set_lock 函数配对使用。

omp_destroy_lock,omp_init_lock 函数的配对操作函数,关闭一个锁。

OpenMP 的子句有以下一些:

private,指定每个线程都有它自己的变量私有副本。

firstprivate,指定每个线程都有它自己的变量私有副本,并且变量要被继承主线程中的初值。

lastprivate,主要是用来指定将线程中的私有变量的值在并行处理结束后复制回主线程中的对应变量。

reduce,用来指定一个或多个变量是私有的,并且在并行处理结束后这些变量要执行指定的运算。

nowait,忽略指定中暗含的等待。

num_threads,指定线程的个数。

schedule,指定如何调度 for 循环迭代。

shared,指定一个或多个变量为多个线程间的共享变量。

ordered,用来指定 for 循环的执行要按顺序执行。

copyprivate,用于 single 指令中的指定变量为多个线程的共享变量。

copyin,用来指定一个 threadprivate 的变量的值要用主线程的值进行初始化。

default,用来指定并行处理区域内的变量的使用方式,缺省是 shared。

2. parallel 指令的用法

parallel 是用来构造一个并行块的,也可以使用其他指令如 for、sections 等和它配合使用。

在 C/C++中，parallel 的使用方法如下：
#pragma omp parallel [for | sections] [子句[子句]…]
{
        //代码
}
parallel 语句后面要跟一个大括号对将要并行执行的代码括起来。
void main(int argc, char * argv[]) {
    #pragma omp parallel
    {
            printf("Hello, World! \n");
    }
}
执行以上代码将会打印出以下结果：
Hello, World!
Hello, World!
Hello, World!
Hello, World!
可以看得出 parallel 语句中的代码被执行了 4 次，说明总共创建了 4 个线程去执行 parallel 语句中的代码。
也可以指定使用多少个线程来执行，需要使用 num_threads 子句：
void main(int argc, char * argv[]) {
    #pragma omp parallel num_threads(8)
    {
            printf("Hello, World!, ThreadId=%d\n", omp_get_thread_num());
    }
}
执行以上代码，将会打印出以下结果：
Hello, World!, ThreadId = 2
Hello, World!, ThreadId = 6
Hello, World!, ThreadId = 4
Hello, World!, ThreadId = 0
Hello, World!, ThreadId = 5
Hello, World!, ThreadId = 7
Hello, World!, ThreadId = 1
Hello, World!, ThreadId = 3
从 ThreadId 的不同可以看出创建了 8 个线程来执行以上代码。所以 parallel 指令是用来为一段代码创建多个线程来执行它的。parallel 块中的每行代码都被多个线程重复执行。
和传统的创建线程函数比起来，相当于为一个线程入口函数重复调用创建线程函数来创建线程并等待线程执行完。

3. for 指令的使用方法

for 指令则是用来将一个 for 循环分配到多个线程中执行。for 指令一般可以和 parallel 指令合起来形成 parallel for 指令使用,也可以单独用在 parallel 语句的并行块中。

#pragma omp [parallel] for [子句]
     for 循环语句

先看看单独使用 for 语句时是什么效果。

```
int j = 0;
#pragma omp for
    for ( j = 0; j < 4; j++ ){
        printf("j = %d, ThreadId = %d\n", j, omp_get_thread_num());
    }
```

执行以上代码后打印出以下结果:

j = 0, ThreadId = 0
j = 1, ThreadId = 0
j = 2, ThreadId = 0
j = 3, ThreadId = 0

从结果可以看出 4 次循环都在一个线程里执行,可见 for 指令要和 parallel 指令结合起来使用才有效果。如以下代码就是 parallel 和 for 一起结合成 parallel for 的形式使用的:

```
int j = 0;
#pragma omp parallel for
    for ( j = 0; j < 4; j++ ){
        printf("j = %d, ThreadId = %d\n", j, omp_get_thread_num());
    }
```

执行后会打印出以下结果:

j = 0, ThreadId = 0
j = 2, ThreadId = 2
j = 1, ThreadId = 1
j = 3, ThreadId = 3

可见,循环被分配到四个不同的线程中执行。

上面这段代码也可以改写成以下形式:

```
int j = 0;
#pragma omp parallel
{
#pragma omp for
    for ( j = 0; j < 4; j++ ){
        printf("j = %d, ThreadId = %d\n", j, omp_get_thread_num());
    }
}
```

执行以上代码会打印出以下结果：
j = 1, ThreadId = 1
j = 3, ThreadId = 3
j = 2, ThreadId = 2
j = 0, ThreadId = 0
在一个 parallel 块中也可以有多个 for 语句，如：
```
int j;
#pragma omp parallel
{
#pragma omp for
    for ( j = 0; j < 100; j++ ){
        ...
    }

#pragma omp for
    for ( j = 0; j < 100; j++ ){
        ...
    }
...
}
```
for 循环语句中，书写是需要按照一定规范来写才可以的，即 for 循环小括号内的语句要按照一定的规范进行书写，for 语句小括号里共有三条语句：

for(i=start; i < end; i++)

i=start; 是 for 循环里的第一条语句，必须写成"变量=初值"的方式。如 i=0

i < end; 是 for 循环里的第二条语句，这个语句里可以写成以下 4 种形式之一：

变量 < 边界值

变量 <= 边界值

变量 > 边界值

变量 >= 边界值

例如 i>10 i< 10　及 i>=10 i>10 等。

最后一条语句 i++ 可以有以下 9 种写法之一：

i++

++i

i--

--i

i += inc

i -= inc

i = i + inc

i = inc + i

i = i - inc

例如 i += 2; i - = 2; i = i + 2; i = i - 2; 都是符合规范的写法。

4. sections 和 section 指令的用法

section 语句是用在 sections 语句里用来将 sections 语句里的代码划分成几个不同的段，每段都并行执行。用法如下：

#pragma omp [parallel] sections [子句]
{
    #pragma omp section
    {
    }
}

先看以下的例子代码：

```
void main(int argc, char *argv)
{
#pragma omp parallel sections {
#pragma omp section
    printf("section 1 ThreadId = %d\n", omp_get_thread_num());
#pragma omp section
    printf("section 2 ThreadId = %d\n", omp_get_thread_num());
#pragma omp section
    printf("section 3 ThreadId = %d\n", omp_get_thread_num());
#pragma omp section
    printf("section 4 ThreadId = %d\n", omp_get_thread_num());
}
```

执行后将打印出以下结果：

section 1 ThreadId = 0

section 2 ThreadId = 2

section 4 ThreadId = 3

section 3 ThreadId = 1

从结果中可以发现第 4 段代码执行比第 3 段代码早，说明各个 section 里的代码都是并行执行的，并且各个 section 被分配到不同的线程执行。

使用 section 语句时，需要注意的是这种方式需要保证各个 section 里的代码执行时间相差不大，否则某个 section 执行时间比其他 section 过长就达不到并行执行的效果了。

上面的代码也可以改写成以下形式：

```
void main(int argc, char *argv)
{
#pragma omp parallel {
```

```
#pragma omp sections
{
#pragma omp section
        printf("section 1 ThreadId = %d\n", omp_get_thread_num());
#pragma omp section
        printf("section 2 ThreadId = %d\n", omp_get_thread_num());
}
#pragma omp sections
{

#pragma omp section
        printf("section 3 ThreadId = %d\n", omp_get_thread_num());
#pragma omp section
        printf("section 4 ThreadId = %d\n", omp_get_thread_num());
}
}
```

执行后将打印出以下结果：
section 1 ThreadId = 0
section 2 ThreadId = 3
section 3 ThreadId = 3
section 4 ThreadId = 1

这种方式和前面那种方式的区别是，两个 sections 语句是串行执行的，即第二个 sections 语句里的代码要等第一个 sections 语句里的代码执行完后才能执行。

用 for 语句来分摊是由系统自动进行，只要每次循环间没有时间上的差距，那么分摊是很均匀的，使用 section 来划分线程是一种手工划分线程的方式，最终并行性的好坏依赖于程序员。

例如，下面是计算图像每一列数据取中值的代码。

```
#pragma omp parallel for private(x,y,buffer)
    for (x=0;x<width;x++)
    {
        for (y=0;y<height;y++)
        {
         buffer[y]=PIX(y,x);
        }
            std::sort(buffer, buffer+height);
            PIX(0,x)=window[height/2];
            TRACE(_T("%d\n"),x);
    }
```

PIX(0,x) 中储存的就是图像每一列数据取中值的结果。

第二节 基于 MPI 与 OpenMP 混合并行计算

MPI(Message Passing Interface)是消息传递并行编程的代表和实施标准,可以轻松地支持分布存储和共享存储拓扑结构;OpenMP 是为共享存储环境编写并行程序而设计的一个应用编程接口,是当前支持共享存储并行编程的工业标准。在多核 PC 机群中,结合 MPI 与 OpenMP 技术,充分利用这两种编程模型的优点,在付出较小的开发代价基础上,尽可能获得较高的性能。

按照在 MPI 进程间消息传递方式和时机,混合模型可以分为以下两种:

(1)单层混合模型(Hybrid Master-only)。MPI 调用发生在应用程序多线程并行区域外,MPI 实现进程间的通信由主线程执行。该混合模型编程易于实现,即在基于 MPI 模型程序的关键计算部分加上 OpenMP 循环命令# pramgma omp parallel 即可。

(2)多层混合模型(Hybrid Multiple)。MPI 调用可以发生在应用程序多线程并行区域内,进程间通信的可由程序任何区域内的任何一个或一些线程完成。在该模型中,当某些线程进行通信时,其他的非通信线程同时进行计算,实现了通信与计算的并行执行,优化了进程间的通信阻塞问题。

1. 单层混合模型(Hybrid Master-only)的实现

MPI_Init_thread( );//MPI 初始化
MPI_Com m_rank( );
MPI_Comm_size( );
MPI_Send( );//MPI 进程间通信
MPI_Recv( );
omp_set_num_threads(n);
#prama omp parallel
{ # pragma omp for
for( ) { //计算部分}}
MPI_Send( );//MPI 进程间通信
MPI_Recv( );
MPI_Finalize( );

2. 多层混合模型(Hybrid Multiple)的实现

MPI_Init _thread( );//MPI 初始化
MPI_Comm_rank( );
MPI_Comm _size( );
pragma omp parallel private(…)
{ # pragma omp for
for(…){//计算部分}
# pragma omp barrier//线程同步
# pragma omp master/ /或者# pragma omp single

{MPI_Send( );//MPI 进程间通信
MPI_Recv( );}
#pragma omp barrier}
MPI_Send( );//MPI 进程间通信
MPI_Recv( );
MPI_Finalize( );

第三节　GPU 并行计算简介及应用

一、GPU 并行计算简介

多核 CPU 和图形处理器(Graphic Processing Unit, GPU)的高速发展，不但促进了图像处理、虚拟现实、计算机仿真等领域的快速发展，同时也为利用 GPU 进行图形处理以外的高性价比绿色通用计算提供了良好的运行平台。因此，GPU 的通用计算已成为高性能计算领域中的热点研究课题之一。伴随着传感器技术的不断进步，人们获取地表信息的手段越来越多样快捷。面对数据源的多样化与数据量的成倍增长，许多常规算法很难满足对海量数据进行高速计算的要求。而现代图形硬件 GPU 日益增加的可编程性和高效能计算能力，则为摄影测量与遥感中图像并行化处理技术的加速提供很大的空间。

GPU 加速计算是利用一个图形处理器（GPU）以及一个 CPU 来加速科学、工程以及企业级应用程序。CPU 与 GPU 之间处理任务的区别在于：CPU 由专为顺序串行处理而优化的几个核心组成；而 GPU 则由数以千计的更小、更高效的核心组成，这些核心专为同时处理多任务而设计。如图 9-1 所示。

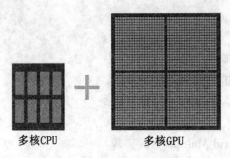

图 9-1　CPU 与 GPU

随着计算从 CPU(中央处理)向 CPU 与 GPU 协同处理方向的发展，NVIDIA 于 2007年发明了统一计算设备架构(Compute Unified Device Architecture, CUDA)，它以 C 语言为基础，可以直接写出在显示芯片上执行的程序，而不需要去学习特定的显示芯片指令或是特殊的结构，是当前比较成熟的通用计算模型构架。利用 GPU 的高度并行性特点将科学计算算法迁移至 CUDA 通用并行计算架构，可以在 GPU 上获得比 CPU 平均数十倍甚至上百倍的性能提升。

CUDA 程序是 GPU 和 CPU 的混合代码，其创建机制是：首先创建 CUDA 程序；然后使用 NVCC 编译器编译程序，EDG 将 GPU 和 CPU 的代码分离，Open64 将 GPU 代码编译成在设备中的 PTX(Parallel Thread eXecution)汇编码或 CUBIN 对象(也就是生成调用 CUDA 驱动的代码)。最后调用链接器把编译好的模块组合在一起，和 CUDA 库与标准 C/C++库链接成为最终的 CUDA 应用程序。值得注意的是，如果 CUBIN 文件支持应用程序把 kernel 发射到 GPU 的架构，则可以使用 CUBIN，否则，CUDA Runtime 将会载入 PTX，同时用动态编译器 JIT(Just In Time compilation，JIT)将 PTX 编译成 CUBIN 格式或者二进制代码，发射到 GPU 上。如果碰到主机代码，则调用主机平台编译器进行编译。

CUDA 的 Grid-Block-Thread 三级线程管理结构是 CUDA 编程模型思想的核心内容。在 CUDA 的架构下，一个程序分为两个部分：host 端和 device 端。Host 端是指在 CPU 上执行的部分，而 device 端则是在显示芯片上执行的部分；Device 端的程序又称为"kernel"。通常 host 端程序会首先将数据准备好复制到显卡的内存中，然后再由显示芯片执行 device 端程序，最后再由 host 端程序将结果从显卡的内存中取回。

在 CUDA 架构下，显示芯片执行时的最小单位是 thread。一方面，一个 block 由多个 thread 组成(包含 thread 数目是有限的)，每个 block 中的 thread 不仅能存取同一块共享内容，且其可以快速进行同步动作；另一方面，执行相同程序的 block 则可以组成 grid 不同的 grid 则可以执行不同的程序，即 kernel。

二、Windows 环境下 CUDA 开发环境的安装及配置

安装 CUDA 前，必须确认以下软件或者硬件已安装：
（1）Windows XP 及以上操作系统；
（2）VS2005 及以上版本；
（3）支持 CUDA 的 Geforce 8x/9x/1xx/2xx/3xx/4xx 系列显卡，或 MCP7x，ION 芯片组。
CUDA 安装具体步骤为：

1. 下载安装文件

在下载地址：https：//developer.nvidia.com/cuda-downloads 获取最新版本的 driver，CUDA toolkit，CUDA SDK。无论显卡的 CUDA 计算能力如何，始终应该使用最新版本的驱动和开发工具(这里以 CUDA6.0 为例)。注意 driver 和 toolkit/SDK 的版本应该与操作系统匹配。

2. 安装文件

依次安装 driver，toolkit 及 SDK。安装完毕后进入 SDK 目录下的 C/bin 子目录下，根据操作系统不同进入 Win32/release 或者 Win64/Release 目录下，运行 devicequery.exe，确认平台上的所有支持 CUDA 设备已被识别，信息正确无误。然后运行 MatrixMul.exe 确认显卡能够分配显存并调用 kernel 进行计算。

3. 配置环境变量

安装完成之后，进行 CUDA 的路径设置,可以在控制台下输入 set cuda 查看现在的环境变量。安装完成之后可以看到 CUDA_PATH 和 CUDA_PATH_V6_0 两个环境变量已经设置成功。接下来在系统环境变量上添加如下的变量：

CUDA_BIN_PATH   %CUDA_PATH%\bin
CUDA_LIB_PATH   %CUDA_PATH%\lib\Win32
CUDA_SDK_BIN   %CUDA_SDK_PATH%\bin\Win32
CUDA_SDK_LIB   %CUDA_SDK_PATH%\common\lib\Win32
CUDA_SDK_PATH   C:\CUDA\CUDA Samples\(注:这个路径需要根据自己的SDK位置进行调整)

重新启动计算机,以使环境变量生效,完成后可以在控制台来运行 C:\NVIDA CUDA\CUDA Samples\Bin\win32\Release 下的 bandwidthTest.exe 和 deviceQuery.exe 来检测。

4. 配置生成规则

将 SDK 目录下 C/common 目录下的 cuda，rules 拷贝到 VS 安装目录下的 VC\VCProjectDefaults 子目录下。建立一个 Win32 工程,并在项目上右击,选择 custombuildrules,在 CUDA build rule 前打勾。在工程中新建一个.cu 文件,右击属性后查看自定义生成规则是否已经是 CUDA build rule。

5. 添加语法高亮

关闭 VS,将 SDK 目录下 C\doc\syntax_highlighting\visual_studio_8 子目录下的 usertype.dat 拷贝到 Microsoft Visual Studio 目录的 \Common7\IDE 子目录下。如果 usertype.dat 已经存在,则将其中的内容添加到已有文件中。打开 visual stuido,选择 "Tools→Options…" 中 "text eidtor" 下的 "file extension",添加.cu,并将其编辑器设为 Microsoft Visual C++,选择 "add" 后,点 "OK",然后打开一个.cu 文件,检查关键字是否已经高亮显示。

具体配置步骤为:

(1) 打开 VS2010 并建立一个空的 Win32 控制台项目。

(2) 右键点击 "源文件 → 添加 → 新建项"。在打开的对话框中选择新建一个 CUDA 格式的源文件(如果只是要调用 CUDA 库编写程序而不需要自行调用核函数分配块,线程的话也可以建立 .cpp 的源文件)。

(3) 右键点击 "工程 → 生成自定义"。在弹出的对话框中勾选 "CUDA 6.0(.targets.props)" 选项。

(4) 右键点击 "项目 → 属性 → 配置属性 → VC++目录",添加以下两个包含目录:

C:\Program Files\NVIDIA GPU Computing Toolkit\CUDA\v6.0\include

C:\ProgramData\NVIDIA Corporation\CUDA Samples\v6.0\common\inc

再添加以下两个库目录:

C:\Program Files\NVIDIA GPU Computing Toolkit\CUDA\v6.0\lib\x64

C:\ProgramData\NVIDIA Corporation\CUDA Samples\v6.0\common\lib\x64

(5) 右键点击 "项目 → 属性 → 配置属性 →连接器 → 常规 → 附加库目录",添加以下目录:

$(CUDA_PATH_V6_0)\lib\ $(Platform)

(6) 右键点击 "项目 → 属性 → 配置属性 →连接器 → 输入 → 附加依赖项",添加以下库:

cublas.lib

cublas_device.lib
cuda.lib
cudadevrt.lib
cudart.lib
cudart_static.lib
cufft.lib
cufftw.lib
curand.lib
cusparse.lib
nppc.lib
nppi.lib
npps.lib
nvblas.lib（32 位系统请勿附加此库！）
nvcuvenc.lib
nvcuvid.lib
OpenCL.lib

（7）右键点击"项目 → 属性",将项类型设置为 CUDA C/C++。
（8）如果操作平台是 64 位,则需打开配置管理器,点击"新建",选择 X64 平台。

三、应用实例

下面给出利用 CUDA 加速遥感影像植被指数计算实例。该程序主要实现了读入一幅图像,对它进行植被指数提取计算,然后输出的功能。代码如下所示:

```
#include "stdio.h"
#include" cuda_runtime.h"
#include "gdal_include/gdal.h"
#include "gdal_include/gdal_priv.h"
#pragma comment(lib,"gdal_i.lib")
using namespace std;
typedef    struct
{
    int x;//左上角在图像中的 X 位置
    int y;//左上角在图像中的 Y 位置
    int w;//分块宽度
    int h;//分块高度
}GSI_Block;//数据分块信息
//kernel 函数
    -global-void function_dev(float * g_r_dev,   float * g_n_dev,   float * g_o_dev,
```

```
double * dm_dev,int w,int h,bool nf)
{
    int m = blockDim.x * blockIdx.x + threadIdx.x;
    while (m < w * h)
    {
        if(nf==0)
        {
            float q=(g_n_dev[m]-g_r_dev[m])/(g_n_dev[m]+g_r_dev[m]);
            g_o_dev[m]=q;
        }
        else
        {
            g_o_dev[m]=(int)255*(g_o_dev[m]-dm_dev[0])/(dm_dev[1]-dm_dev[0]);
        }
        m+=256*384;
    }
}
//并行处理函数
void paralfunc(float *g_r,  float *g_n,  float *g_o,double * dm,int w,int h,bool nf)
{
    if(nf==0)
    {
    float *g_r_dev;
    cudaError(cudaMalloc((void**)&g_r_dev,w*h*sizeof(float)));
    cudaError(cudaMemcpy(g_r_dev,g_r,w*h*sizeof(float),cudaMemcpyHostToDevice));
    float *g_n_dev;
    cudaError(cudaMalloc((void**)&g_n_dev,w*h*sizeof(float)));
    cudaError(cudaMemcpy(g_n_dev,g_n,w*h*sizeof(float),cudaMemcpyHostToDevice));
    float *g_o_dev;
    cudaError(cudaMalloc((void**)&g_o_dev,w*h*sizeof(float)));
        function_dev<<<384,256>>>(g_r_dev,  g_n_dev,  g_o_dev,NULL, w, h, nf);
        cudaError(cudaMemcpy(g_o,g_o_dev,w*h*sizeof(float),cudaMemcpyDeviceToHost));
    cudaFree(g_r_dev);
```

```
            cudaFree(g_n_dev);
            cudaFree(g_o_dev);
        }
        else
        {
            float * g_o_dev;
            cudaError(cudaMalloc((void * * )&g_o_dev,w * h * sizeof(float)));
            cudaError(cudaMemcpy(g_o_dev, g_o, w * h * sizeof(float), cudaMemcpy
HostToDevice));
            double * dm_dev;
            cudaError(cudaMalloc((void * * )&dm_dev,2 * sizeof(double)));
            cudaError(cudaMemcpy(dm_dev, dm, 2 * sizeof(double), cudaMemcpy
HostToDevice));
            function_dev<<<384,256>>>(NULL, NULL, g_o_dev,dm_dev, w, h, nf);
            cudaError(cudaMemcpy(g_o, g_o_dev, w * h * sizeof(float), cudaMemcpy
DeviceToHost));
            cudaFree(g_o_dev);
            cudaFree(dm_dev);
        }
    }
//分块函数
void GetBlockNums(int lWidth, int lHeight, GSI_Block * vBlock, int nBlockSize)
{
        GSI_Block blocktemp;
        int j=0;
        for (int i = 0; i < lHeight; i += nBlockSize)
        {
            blocktemp.x = 0;blocktemp.y = i;
            blocktemp.w = lWidth;
            blocktemp.h = min(nBlockSize, lHeight-i);
            vBlock[j++]=blocktemp;
        }
    }
int main(int argc, char * argv[])
{
        //初始化
        GDALAllRegister();
        GSI_Block vBlock[100];
```

```cpp
int i,j,nw,nh;
GDALDataset * p;
GDALDataset * r;
p=(GDALDataset * )GDALOpen("mul-1024.tif",GA_ReadOnly);
int ncount=p->GetRasterCount();
GDALRasterBand * * pand;
GDALRasterBand * * rand;
rand=new GDALRasterBand * [ncount];
pand=new GDALRasterBand * [1];
for(i=0;i<ncount;i++)
{
    pand[i]=p->GetRasterBand(i+1);
}
nw=pand[0]->GetXSize();
nh=pand[0]->GetYSize();
GetBlockNums(nw, nh, vBlock, 256 );
float * gray_in;
GDALDriver * dri;
char * * papszOptions2 = NULL ;
const char * pszFormat = "GTiff";
dri=GetGDALDriverManager()->GetDriverByName(pszFormat);
r =(GDALDataset * ) dri->Create("ndvi_cuda.tif",nw,nh,1,GDT_Float32,papszOptions2);
rand[0]=r->GetRasterBand(1);
double dm[2] = {0.0,255.0};
//分块处理
for(i=0;i<100;i++)
{
    if(vBlock[i].w==nw)
    {
        int m,n;
        int bw=vBlock[i].w;int bh=vBlock[i].h;
        gray_in=new float[2 * bw * bh];
        float * gray_out=new float[bw * bh];
        //影像读取
        GDALRasterIO(pand[2],GF_Read,vBlock[i].x,vBlock[i].y,bw,bh,gray_in,bw,bh,GDT_Float32,0,0);
        GDALRasterIO(pand[3],GF_Read,vBlock[i].x,vBlock[i].y,bw,bh,&gray_in[bw
```

```
        * bh],bw,bh,GDT_Float32,0,0);
    //执行并行处理函数
    paralfunc(gray_in,&gray_in[bw*bh],gray_out,NULL,bw,bh,0);
    //影像写入
    GDALRasterIO(rand[0],GF_Write,vBlock[i].x,vBlock[i].y,bw,bh,gray_out,bw,bh,
GDT_Float32,0,0);
            delete []gray_in;
            delete []gray_out;
        }
    }
    //最大最小值统计
    GDALComputeRasterMinMax(rand[0],TRUE,dm);
    //分块处理
    for(i=0;i<100;i++)
    {
        if(vBlock[i].w==nw)
        {
        int m,n;
        int bw=vBlock[i].w;int bh=vBlock[i].h;
        float* gray_out=new float[bw*bh];
    //影像读取
    GDALRasterIO(rand[0],GF_Read,vBlock[i].x,vBlock[i].y,bw,bh,gray_out,bw,bh,
GDT_Float32,0,0);
    //执行并行处理函数
    paralfunc(NULL,NULL,gray_out,dm,bw,bh,1);
    //影像写入
    GDALRasterIO(rand[0],GF_Write,vBlock[i].x,vBlock[i].y,bw,bh,gray_out,bw,bh,
GDT_Float32,0,0);
            delete []gray_out;
        }
    }
    GDALClose(p);
    GDALClose(r);
    return 0;
}
```

第十章　OpenRS 简介与应用

第一节　OpenRS 简介与安装

一、OpenRS 简介

开放式遥感数据处理与服务平台 OpenRS 是武汉大学测绘遥感信息工程国家重点实验室江万寿研究员负责的 863 重点项目。目的是建立开放式的遥感数据处理和应用服务的体系架构，形成遥感公共数据产品加工标准规范，实现高性能的遥感数据快速处理，最终研制开发多源遥感数据处理软件和遥感信息产品加工系统，为"多源遥感数据综合处理与服务系统"提供技术支撑，提升我国多源遥感数据数据处理技术与应用水平。

OpenRS 是一种"平台+插件"的架构，其中，平台负责插件的扫描、对象的注册、查询和创建。它提供了最常用的图像读写、显示、漫游等基本图像处理与分析功能以及摄影测量中的传感器成像模型、地理坐标管理等。为平台设计插件时，不需再考虑设计和实现平台的这些功能，因而能避免重复编写代码，有助于提高研发的效率和质量。

OpenRS 的总体技术路线是基于插件技术和网络服务技术建立的开放式、支持分布式并行处理的遥感数据处理平台。如图 10-1 所示，OpenRS 软件平台可以粗略地分为基础支撑层、处理插件层和系统应用层。

基础支撑层提供与应用无关的插件管理、并行任务管理与执行、流程定制与执行等基础功能。这部分基础功能一般是与遥感应用无关的通用功能。而数据管理和显示则与遥感影像有关。为了简化起见，作为遥感数据处理平台，我们把数据管理与显示也作为基础支撑层的一部分。处理插件层是与应用有关的业务处理功能实现层，该层强调算法的扩展性和替换性。通过插件系统的支持，可以通过插件机制动态地增加系统的处理功能和处理能力。系统应用层是直接面向用户的应用系统，包括用户界面和应用逻辑。可以通过流程定制管理器组合处理插件提供的处理功能，形成面向某一具体应用的完整处理链；也可以直接在程序中按编码的形式固化处理链条。在任务的执行上，应用系统可以在本地调用处理节点的功能，也可以把处理任务提交给分布式任务管理器，由分布式任务管理器负责把任务分解分派到集群服务器的处理节点，完成任务的并行处理。该平台架构提供开放式的平台开发环境和分布式处理的架构，但处理算法的质量还需要算法本身和外部数据的质量来保证。

开放式遥感平台采用全插件的系统结构。二次开发人员和平台开发人员一样可以在算法级(包括数据源、几何模型等)、执行模块级、界面级等多个层次对系统功能进行动态

第十章 OpenRS 简介与应用

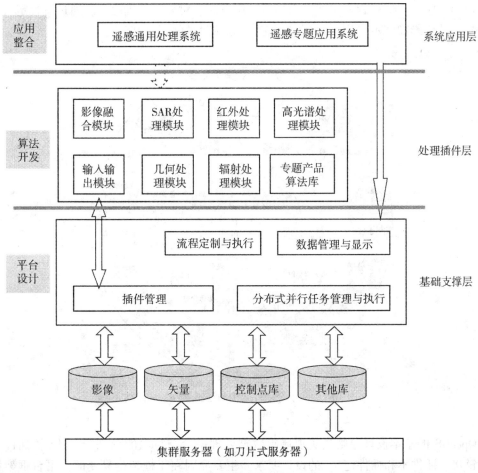

图 10-1 总体技术路线示意图

扩展。系统采用类似 Eclipse 的微内核结构。如图 10-2 所示，系统分为插件系统层、服务插件层、对象插件层、应用程序层四个层次：

内核为插件管理系统，提供注册服务、日志服务、错误处理服务等三大基础服务。这里的服务概念是指在整个 OpenRS 运行生命周期内唯一的对象，可用于创建其他对象或其他信息的输出（这里的"服务"和 SOA 无关）。插件系统与具体应用无关，相当于"软件总线"，是一个通用的架构。

服务插件层：提供可替代的系统服务或领域相关的模块级服务功能。可替代的系统服务如属性的序列化、流程的解析、界面的扩展、分布式并行处理等。遥感数据处理相关的服务如空间参考、影像处理、矢量处理、图层管理等基础服务。通过服务插件层，系统从一个通用的插件系统扩展服务于某一具体应用领域的基本架构，具备了搭建具体应用程序的基础。服务层的对象是一个单例对象，无需应用程序创建。

对象插件层：提供可创建的多例对象，用于实现文件读取、传感器几何模型、影像变换、影像分割、影像聚类、影像分类等各种算法对象和其他对象。

应用程序层：应用程序构建于内核层、服务层、对象层基础之上，通过流程定义、算法配置，结合影像演示等用户界面实现完整的遥感数据处理功能。

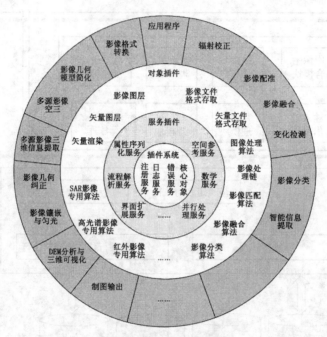

图 10-2　基于全插件架构的遥感软件层次结构

OpenRS 的基本设计原则是实现遥感数据处理平台与算法的松耦合，达到"插件不和平台链接、插件不和插件连接"的设计要求。主要原因在于框架是易变的，平台框架始终都将经历不断发展演化的过程，需要逐步得到完善。框架是业务流，可复用性相对更低。而插件是功能模块，插件模块设计时应忽略框架的存在。从软件设计的角度看，目前面向对象设计已经发展为组件技术。面向对象技术的基础是封装，即接口与实现分离；面向对象的核心是多态，这是接口和实现分离的更高级升华，使得在运行时可以动态地根据条件来选择隐藏在接口后面的实现，面向对象的表现形式是类和继承。面向组件技术建立在对象技术之上，它是对象技术的进一步发展，类概念仍然是组件技术中一个基础的概念，但是组件技术更核心的概念是接口。组件技术的主要目标是粗粒度复用，即组件的复用，如一个 dll、一个中间件，甚至一个框架。一个组件可以由一个类或多个类及其他元素(枚举、)组成，但是组件有个很明显的特征，就是它是一个独立的物理单元，经常以非源码的形式(如二进制，IL)存在。一个完整的组件中一般有一个主类，而其他的类和元素都是为了支持该主类的功能实现而存在的。为了支持这种物理独立性和粗粒度的复用，组件需要更高级的概念支撑，其中最基本的就是属性和事件。在 OpenRS 中，组建技术进一步发展为插件技术。插件就是一种特殊的组件。当然考虑到平台无关性，OpenRS 的插件是和 Window 组件无关的特殊动态库。总之，接口、属性、事件是 OpenRS 面向对象设计的三大基石。

影像链的思路是影像链中的对象具有相同的接口，这样对于影像处理算法和显示程序只需要针对"影像源"接口编程，而不需要考虑输入的"影像源"是原始的影像数据块还是处理算法动态生成的影像数据块，如图 10-3 所示。影像链模式具有如下优点：多个处理过程可以动态链接、多个处理过程可以任意组合。图像块的读入、合并、重投影、图像滤波和输出等一系列过程可以在线组合。图像显示可以插入图像链的任意位置，用于查看图像处理的当前状态。

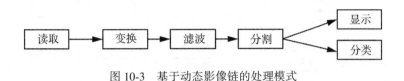

图 10-3 基于动态影像链的处理模式

二、OpenRS 集成环境

1. 主控制模块

OpenRS 主控模块主要的功能是实现 OpenRS 相关软件模块的快速集成。在功能上模仿 ERDAS，在实现上模仿 Windows 桌面。如图 10-4 所示，OpenRS 主控模块的图标其实就是执行程序快捷方式。快捷方式放在 OpenRS 执行程序目录下的 iconPanel 子目录中。

图 10-4 OpenRS 主控模块

除了集成执行程序外，OpenRS 主控模块还提供了对象查询、插件查询、RDF 查询功能。如图 10-5 所示，对象查询按对象类别树列出所有的对象。对于每个对象，可以查询对象所在的插件。如果不知道一个对象的 ID，可以在对象树上查找到该对象，然后拷贝出该对象的 ID。

2. 对象执行器 orsExeRunner

orsExeRunner 专门用于查询和执行可执行对象的，orsExeRunner 的界面如图 10-6 所示。如果说 OpenRS 的插件对象查询能查询所有的 OpenRS 对象的话，那么 orsExeRunner 可以查询和执行所有的可执行对象。可以指定运行时指派的 CPU 线程数、可以指定在本地执行还是远程执行、指定并行执行场景对象。WriteWSDL 按钮则提供一键式服务包装功能，生成用于 IIS 的 C#代码或用于 OpenRS 专用的服务器的 C 代码。

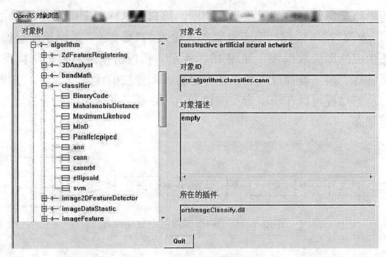

图 10-5　对象查询界面

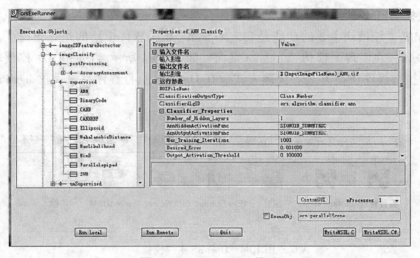

图 10-6　orsExeRunner 界面

3. 综合显示与集成环境 orsViewer

orsViewer 模块为 OpenRS 主要处理和显示界面，包括图层管理窗口、图像显示窗口、工作流窗口和矢量工具，如图 10-7 所示。orsViewer 的设计目标是能够融合 ArcMap、ERDAS、ENVI 等主流 GIS、遥感软件的相关功能，形成一个具有较强栅格、矢量显示、能够动态集成影像处理功能的综合处理环境。orsViewer 集成了 orsExeRunner 中的所有算法对象。

三、OpenRS 开源代码下载编译

下载 OpenRS 开源代码之前，首先需要下载 TortoiseSVN 管理软件（下载地址为

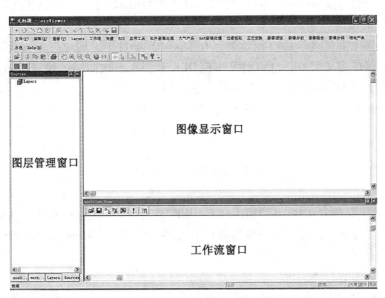

图 10-7 orsViewer 界面

http：//tortoisesvn.net/downloads），然后在磁盘根目录下新建"OpenRS"目录，使用 svn 下载源代码到此目录下，如磁盘为 E 盘，则在新建的"E：\ OpenRS"文件夹上右击，弹出 svn 的菜单，选择 updata，在弹出的下载对话框中输入"https：//202.114.114.15：8443/svn/OpenRS/trunk/"，一般客户用账户"guest"，口令为空，即可进行下载。下载得到的文件包含两个文件夹"desktop"与"externel"，"externel"文件夹里包含第三方库的源代码压缩包与已经编译好的动态库与静态库（如 GDAL 开源库），"desktop"文件夹中为 OpenRS 工程的文件，其中工作空间文件为"desktop \ build \ vc60 \ openRS.dsw"。在正式编译 OpenRS 平台之前，需要对 OpenRS 进行添加头文件和配置平台环境。具体步骤如下：

（1）添加头文件：首先解压第三方库源文件，将"externel/zip"中所有的压缩包都解压放在"external"文件夹下，然后将"external"目录下的第三方库添加到 VC 的"Option"目录下。

（2）环境变量配置：以 VC6 为例说明，点击"tools→options→directories"，在"lib files"中添加路径"E：\ OpenRS \ external \ lib"；点击"我的电脑→属性→高级→环境变量"，找到 path 项，添加"E：\ OpenRS \ external \ bin"，再重启计算机，即可激活此环境变量。

第二节 OpenRS 应用示例

一、OpenRS 的插件设计与实现

在 OpenRS 平台中，插件的实现由接口类及实现类组成，其中实现类继承自接口类。

本章以新建一个名为 testplugin 的插件为例,讲述在 OpenRS 平台中的具体实现过程。

1. 接口类的设计

接口类作为一个普通的 C++纯虚类,在 OpenRS 平台中,通过继承 orsIObject 或其派生类,并加上接口定义宏 ORS_INTERFACE_DEF 来实现。我们定义接口类名称为 orsISE_testplugin,具体实现代码如下:

```
class orsISE_testplugin: public orsISimpleExe
{
public:
    ORS_INTERFACE_DEF(orsISimpleExe, "testplugin");
};
```

接口 ID 的定义通过 ORS_INTERFACE_DEF 宏实现,并且实现 orsISE_testplugin 类对 orsISimpleExe 类的继承。宏中第一项为继承的父类,第二项为 orsISE_testplugin 接口的 ID。

2. 实现类的设计

实现类是算法的具体实现,它是一个基本的 C++类,通过继承接口来实现接口的相关方法。在 OpenRS 中,系统通过提供类 orsObjectBase 和系列宏 ORS_OBJECT_IMPn(n 为 1,2,3)来帮助实现,宏 ORS_OBJECT_IMPn 中 n 代表要能够被查询到的接口个数。

ORS_OBJECT_IMP3(orsExetestplugin, orsISE_testplugin, orsISimpleExe, orsIExecute, "testplugin", "testplugin");

宏中第一项到第四项分别为继承关系,即 orsExetestplugin 为实现类名称,继承于 orsISE_testplugin,orsISE_testplugin 又继承于 orsISimpleExe,orsISimpleExe 继承于 orsIExecute,后两项为插件信息描述。

3. 插件的注册

在完成接口类和实现类的设计之后,接下来我们就要进行插件的注册了。为了便于调用与版权问题,插件需要提供自身的元信息,包括编写者、版本、插件名称、插件 ID 等。为此,OpenRS 系统提供了一个 orsIPlugin 接口,让插件编写者定义 orsXPlugin 类继承于 orsIPlugin 来填充元信息。具体过程如下:

首先对本例实现类编写创建函数,实现类 orsExetestplugin 的创建函数如下:

```
orsIObject * createExetestplugin( bool bForRegister)
{
    return new orsExetestplugin( bForRegister );
}
```

其次在 orsXPlugin 类里面的 Initialize 函数中,将创建函数的函数指针进行注册:

pRegister->registerObject(createExetestplugin);

其中 pRegister 为获取的平台影像服务指针。

最后通过 ORS_REGISTER_PLUGIN(orsXPlugin)实现融合插件的注册。

通过以上步骤,实现了 testplugin 插件的制作,OpenRS 平台中通过插件的唯一 ID 来进行扫描并加载插件。具体代码如下:AddMenuByExeClass(frameWnd, _T("testplugin"));其中第一项为框架窗口接口类指针,第二项为 testplugin 插件的 ID。

二、融合插件示例

下面介绍基于 OpenRS 设计的融合插件。主要包括 IHS 融合、PCA 融合、乘积融合以及 Brovey 融合等功能。

1. IHS 融合

功能描述:对输入的全色影像和多光谱影像进行 IHS 融合,IHS 融合操作界面如图 10-8 所示。

输入影像:输入多光谱影像和全色影像。

输入参数:多光谱波段号,全色波段号。

如图 10-8 所示。

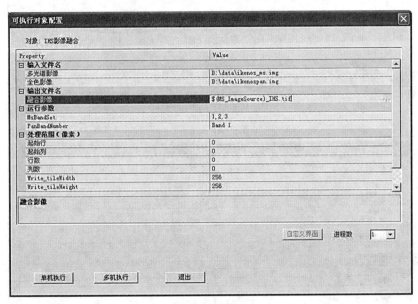

图 10-8 HIS 影像融合

2. PCA 融合

功能描述:对输入的全色影像和多光谱影像进行 PCA 融合,PCA 融合操作界面如图 10-9所示。

输入影像:输入多光谱影像和全色影像。

输入参数:多光谱波段号,全色波段号。

如图 10-9 所示。

3. Brovey 融合

功能描述:对输入的全色影像和多光谱影像进行 PCA 融合,PCA 融合操作界面如图 10-11 所示。

输入影像:输入多光谱影像和全色影像。

输入参数:多光谱波段号,全色波段号。

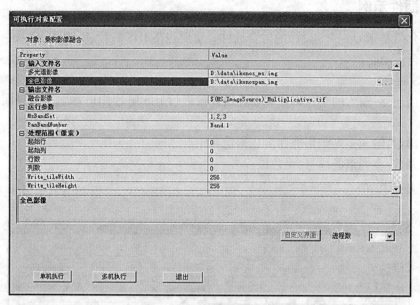

图 10-9 PCA 影像融合

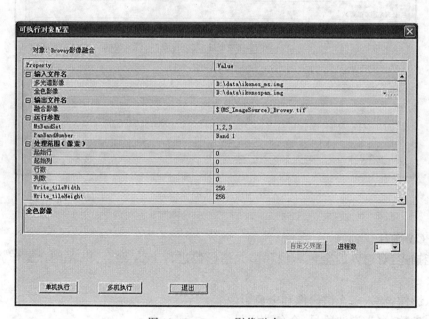

图 10-10 Brovey 影像融合

4. 乘积融合

功能描述:对输入的全色影像和多光谱影像进行乘积融合,乘积融合操作界面如图 11 所示。

输入影像:输入多光谱影像和全色影像。

输入参数:多光谱波段号,全色波段号。

如图 10-11 所示。

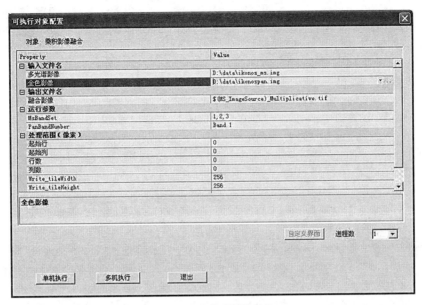

图 10-11 乘积影像融合

第三部分 综合设计

第十一章 基础实习(必做)

实习一 实现 RAW→BMP 格式的转换

【实习目的】

学生通过编程实现 RAW→BMP 格式的转换,掌握 RAW 和 BMP 文件格式特点及其存取方法,为读取各种格式文件、相互转换和后续进行图像处理打下基础。

【实习内容】

根据指导老师的要求,在 VC 编程环境下,用面向对象的程序设计方法,对所给图像实现 RAW 格式图像文件的读取,并将其转换为 BMP 格式文件存储。

【预备知识】

1.熟悉 VC 编程。

2.熟悉 BMP 格式文件结构。

3.读取 BMP 文件的函数原型。

HDIB ReadDIBFile(CFile& file)函数的参数 Cfile &file 为文件名;该函数的返回值是文件中的图像数据的句柄。图 11-1 所示是读取 BMP 文件的流程图。

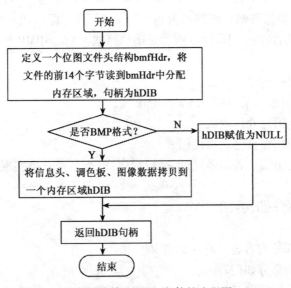

图 11-1 读取 BMP 文件的流程图

4. 图像数据存储为 BMP 文件。

函数原型 BOOL WINAPI SaveDIB(HDIB hDib, CFile& file)的第一个参数 HDIB 表示保存图像数据的内存区域的句柄,第二个参数 CFile& file 为保存图像的文件名。图 11-2 所示是图像数据存储为.bmp 的流程图。

图 11-2　图像数据存储为.bmp 的流程图

【实习原理】

将 RAW 格式文件读到内存,根据 BMP 文件格式给新文件创建四部分内容:BITMAPFILEHEADER(位图文件头)、BITMAPINFOHEADER(位图信息头)、Palette(调色板)、DIB Pixels(DIB 图像数据),将四部分内容写入新文件,即生成一个 BMP 文件,从而实现 RAW 到 BMP 格式的转换。图 11-3 所示是图像格式 RAW→BMP 转换的流程图。

【实习步骤】

1. 打开自己实习一所建的项目。
2. 如图 11-4 所示添加菜单项"RAW→BMP"。
3. 为该菜单建立消息处理函数。
4. 在函数定义处添加自己的源代码。
5. 编译检查语法错误。若编译通过,运行程序,检查是否正确实现 RAW→BMP 格式的转换。

实习完毕后,提交一份实习报告。

【思考题】

1. BMP 文件四部分内容各具有什么意义?
2. BMP 文件第四部分如何写入?
3. 分析所给函数 LPSTR WINAPI FindDIBBits(LPSTR lpbi)的功能。
4. 编程实现 BMP→RAW 格式的转换。
5. 在熟悉 PCX 图像格式的基础上,编程实现 PCX 图像格式的读写。

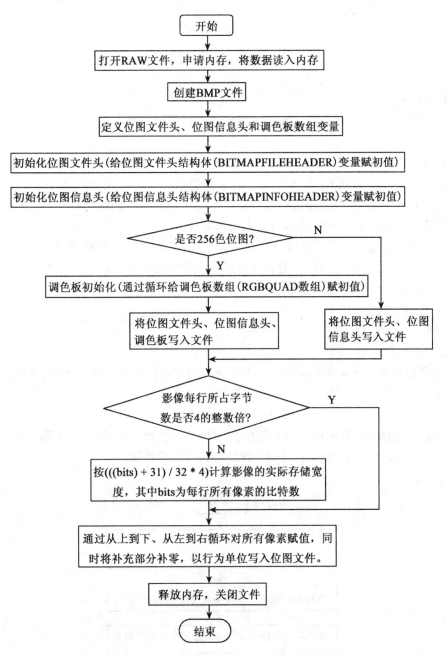

图 11-3　图像格式 RAW→BMP 转换的流程图

图 11-4　添加菜单项

实习二 灰度图像对比度增强

【实习目的】

熟悉图像增强点运算的方法,编程实现图像对比度增强的算法,掌握灰度图像基本处理技术。

【实习内容】

1.在上次实习的基础上,用面向对象的编程思想创建一个新类(可采用 CImgProcess、CBmp 或 CDIB 等与图像处理相关的类名),该类用于实现对位图的各种操作,如读写、信息获取以及各种处理。将上次实习内容加入该新类。

2.认真阅读图像显示程序段以及像素灰度值的读取、改变与存储,按指导老师的要求,编写灰度图像对比度增强的程序,并将灰度图像对比度增强算法加入新类。

【预备知识】

1.熟悉 BMP 位图文件的读取和显示;

2.熟悉 VC 编程中对话框的设计和操作。

【实习原理】

灰度图像对比度增强就是指对图像中所有点的灰度按照某一线性变换函数进行变换。线性变换方程如下:

$$D_0 = f(D_i) = aD_i + b \tag{1}$$

式中,参数 D_i 为输入图像的像素的灰度值,参数 D_0 为输出图像的灰度,a 和 b 由给定条件确定。图 11-5 所示是对一幅图像进行灰度线性拉伸的流程图。

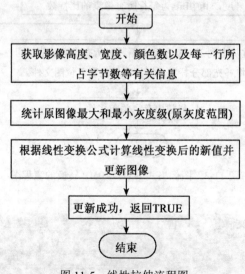

图 11-5 线性拉伸流程图

【实习步骤】

1. 打开自己上次实习所建的项目,为该工程添加一个新类,用于实现对 BMP 格式图像文件进行处理的操作。

2. 在新类中加入相应属性和操作,比如上次的格式转换、读写操作以及这次实习要实现的灰度线性拉伸等操作。

3. 在自己创建的类中添加线性拉伸操作。

4. 如图 11-6 所示添加一栏主菜单"图像增强",再在其下拉菜单中加入一项"线性拉伸"子菜单。

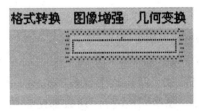

图 11-6　添加菜单项

5. 为子菜单"线性拉伸"建立消息处理函数。

6. 在函数定义处添加自己的源代码(通过自己建的类对象来实现)。

7. 编译检查语法错误。若编译通过,运行程序,检查设计的灰度线性变换程序是否正确。

实习完毕后,提交一份实习报告。

【思考题】

1.图像对比度增强应用在哪些场合?编程实现灰度图像二值化的功能。

2.编程实现灰度图像分段线性变换的功能。

实习三　图像局部处理

【实习目的】

在熟悉图像局部处理方法基础上,编程实现灰度图像局部处理的算法,为提高学生图像处理与分析能力奠定基础。

【实习内容】

熟悉局部处理的卷积算法,按指导老师的要求,在前次实习的基础上完成用 3×3 卷积核实现局部处理的编程;采用合理方式解决卷积运算中特殊像素的输出问题。

【预备知识】

1.熟悉 BMP 位图文件的读取和显示。

2.熟悉 VC 编程中对话框的设计和操作。

【实习原理】

局部处理是指在处理图像某一像素时,其输出值由当前像素邻域的某种变换得到的。

当前像素的邻域一般以当前像素为中心的二维矩阵表示,该矩阵的大小为奇数。

图像空间域平滑、空间域锐化等都属于局部处理。大部分的局部处理都是采用卷积算法来实现。

卷积可以简单地看成加权求和的过程。卷积时使用一个较小的权矩阵,这种权矩阵叫做卷积核。矩阵的大小往往是奇数,而且与邻域大小相同。邻域中的每个像素灰度值与卷积核中对应的元素相乘,所有乘积之和即为邻域中心像素的输出值。比如,对于一个 3×3 的邻域 p 与卷积核 k 卷积后,中心像素 p_5 的输出为:

$$p'_5 = \sum_{i=1}^{9} p_i \cdot k_i \tag{1}$$

其中,$p = \begin{bmatrix} p_1 & p_2 & p_3 \\ p_4 & p_5 & p_6 \\ p_7 & p_8 & p_9 \end{bmatrix}$,$k = \begin{bmatrix} k_1 & k_2 & k_3 \\ k_4 & k_5 & k_6 \\ k_7 & k_8 & k_9 \end{bmatrix}$

卷积核中各元素值叫做权系数。卷积核中权系数决定了图像卷积处理的效果。通常采用 3×3、5×5 或 7×7 的卷积核,一般卷积核的行、列数都是奇数。图 11-7 所示是图像局部处理的流程图。

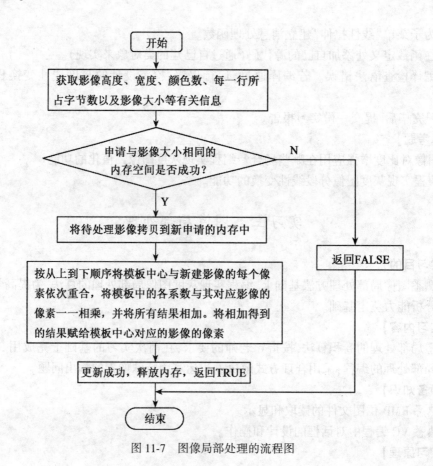

图 11-7 图像局部处理的流程图

【实习步骤】
1. 打开上次实习所建的项目。
2. 在自己创建的类中添加滤波操作。
3. 如图 11-8 所示添加菜单项"3×3 低通滤波"、"3×3 高通滤波"。

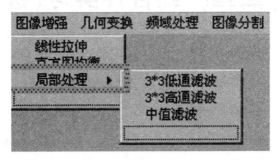

图 11-8　添加菜单项

4. 为新加的两个菜单项分别建立消息处理函数。
5. 在函数体内添加实现 3×3 低通滤波(或 3×3 高通滤波)的源代码。
6. 编译检查语法错误,若编译通过,运行程序,检查是否正确实现局部处理算法。
实习完毕后,提交一份实习报告。

【思考题】
1. 运算过程中图像边界像素如何处理?
2. 卷积计算过程中使用的是变换前的像素还是变换后的像素,该如何解决数据读取问题?
3. 修改程序,实现局部处理的通用算法。

第十二章 综合性实习(选做)

实习一 灰度图像中值滤波

【实习目的】

熟悉中值滤波原理和快速排序的方法,编程实现灰度图像中值滤波功能,探讨其去噪特性及其适用性。提高学生图像增强处理的技能。

【实习内容】

按指导老师的要求,编程实现灰度图像中值滤波功能;针对不同噪声灰度图像,探讨如何应用中值滤波去噪效果最佳。

【预备知识】

1.熟悉图像局部处理的方法。

2.熟悉 BMP 位图文件的读写和显示。

【实习原理】

中值滤波是一种典型的低通滤波器,它的目的是去除噪声,同时能保护图像边缘。中值滤波一般采用一个含有奇数个点的滑动窗口,将窗口中各点灰度值的中值来代替指定点(一般为窗口中心)的灰度值。对于奇数个元素,中值是取窗口中各元素按由小到大排序后中间的灰度值;对于偶数个元素,一般取排序后中间两个元素灰度值的平均值为中值。图 12-1 所示是中值滤波处理的流程图。

【实习步骤】

1.打开上次实习所用的项目。

2.在自己创建的类中添加中值滤波操作。

3.添加如图 12-2 所示的菜单项"中值滤波"。

4.为新加的菜单项建立消息处理函数。

5.在函数体内添加实现中值滤波的源代码。

6.编译检查语法错误。若编译通过,运行程序,检查是否正确实现中值滤波算法。

实习完毕后,提交一份实习报告。

【思考题】

1. 中值滤波处理效果与其他低通滤波器有何不同?

2. 探讨中值滤波窗口大小和形状的选取对滤波效果的影响。

3. 对于图像的不同区域,滤波算子相应地有所不同。若权值选取依赖于区域的灰度中值,且当某点的灰度越接近灰度中值,其权值就相应的越大。试编程实现《中国图像图形学

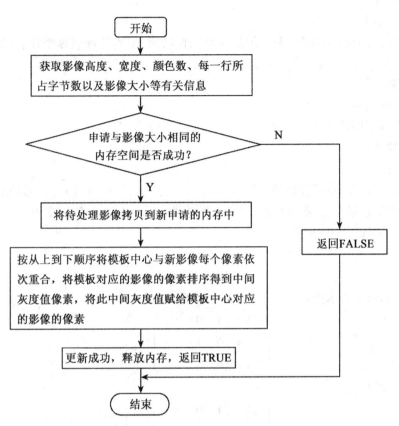

图 12-1 中值滤波处理的流程图

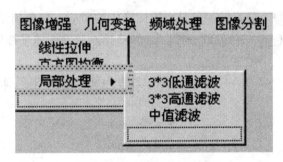

图 12-2 "中值滤波"菜单

报》2004,9(4):408-511 提出的自适应中值滤波算法,并分析其性能。

实习二 图像几何处理:图像平移、缩放和旋转变换

【实习目的】

熟悉图像的几何变换原理和方法,编程实现图像的几种基本几何变换算法如平移、缩

放、旋转变换等。

【实习内容】

在新建的类中加入图像平移、缩放、旋转操作功能,编程实现图像平移、缩放、旋转等变换。

【预备知识】

1. 熟悉面向对象程序设计。
2. 熟悉灰度图像的读写和显示。

【实习原理】

1. 图像平移

将图像中所有的点都按照指定的平移量水平、垂直移动。设(x_0,y_0)为原图像上的一点,图像水平平移量为t_x,垂直平移量为t_y,则平移后点(x_0,y_0)的坐标变为(x_1,y_1)。
(x_0,y_0)与(x_1,y_1)之间的关系为:

$$\begin{cases} x_1 = x_0 + t_x \\ y_1 = y_0 + t_y \end{cases}$$

以矩阵的形式表示为:

$$\begin{pmatrix} x_1 \\ y_1 \\ 1 \end{pmatrix} = \begin{pmatrix} 1 & 0 & +t_x \\ 0 & 1 & +t_y \\ 0 & 0 & 1 \end{pmatrix} \begin{pmatrix} x_0 \\ y_0 \\ 1 \end{pmatrix} \tag{1}$$

它的逆变换:

$$\begin{pmatrix} x_0 \\ y_0 \\ 1 \end{pmatrix} = \begin{pmatrix} 1 & 0 & -t_x \\ 0 & 1 & -t_y \\ 0 & 0 & 1 \end{pmatrix} \begin{pmatrix} x_1 \\ y_1 \\ 1 \end{pmatrix} \tag{2}$$

平移后的图像中每个像素的颜色是由原图像中的对应点颜色确定的。如新图中的(0,0)点的颜色和原图中$(-t_x,-t_y)$处的一样。图12-3所示是对图像平移处理的流程图。

2. 图像旋转

通常是以图像的中心为圆心旋转,按顺时针方向旋转后的图像,如图12-4所示。
旋转前:

$$x_0 = r \cdot \cos b \tag{3}$$
$$y_0 = r \cdot \sin b$$

旋转a角度后:

$$x_1 = r \cdot \cos(b-a) = r \cdot \cos b \cdot \cos a + r \cdot \sin b \cdot \sin a = x_0 \cdot \cos a + y_0 \cdot \sin a$$
$$y_1 = r \cdot \sin(b-a) = r \cdot \sin b \cdot \cos a - r \cdot \cos b \cdot \sin a = -x_0 \cdot \sin a + y_0 \cdot \cos a; \tag{4}$$

以矩阵的形式表示为:

$$\begin{bmatrix} x_1 \\ y_1 \\ 1 \end{bmatrix} = \begin{bmatrix} \cos a & \sin a & 0 \\ -\sin a & \cos a & 0 \\ 0 & 0 & 1 \end{bmatrix} \begin{bmatrix} x_0 \\ y_0 \\ 0 \end{bmatrix} \tag{5}$$

式(5)中,坐标系是以图像的中心为原点,向右为x轴正方向,向上为y轴正方向。它和

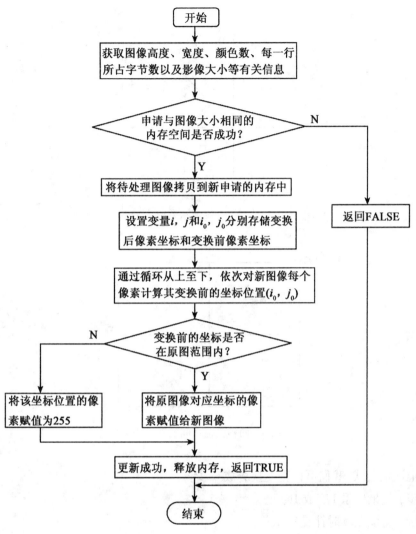

图 12-3 图像平移处理的流程图

以图像左上角为原点,向右为 x 轴正方向,向下为 y 轴正方向的坐标系之间的转换关系如何呢? 如图 12-5 所示。

设图像的宽为 w,高为 h,容易得到:

$$\begin{bmatrix} x_{\mathrm{I}} \\ y_{\mathrm{I}} \\ 1 \end{bmatrix} = \begin{bmatrix} 1 & 0 & 0.5w \\ 0 & -1 & 0.5h \\ 0 & 0 & 1 \end{bmatrix} \begin{bmatrix} x_{\mathrm{II}} \\ y_{\mathrm{II}} \\ 1 \end{bmatrix} \tag{6}$$

逆变换为:

$$\begin{bmatrix} x_{\mathrm{II}} \\ y_{\mathrm{II}} \\ 1 \end{bmatrix} = \begin{bmatrix} 1 & 0 & -0.5w \\ 0 & -1 & 0.5h \\ 0 & 0 & 1 \end{bmatrix} \begin{bmatrix} x_{\mathrm{I}} \\ y_{\mathrm{I}} \\ 1 \end{bmatrix} \tag{7}$$

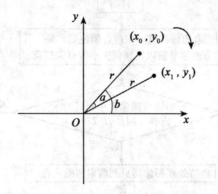

图 12-4 旋转示意图

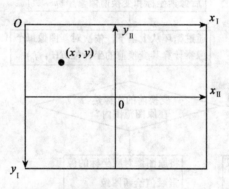

图 12-5 两种坐标系间的转换关系

有了上面的公式,我们可以把变换分成三步:

第一步,将坐标系 I 变成 II;

第二步,将该点顺时针旋转 a 角;

第三步,将坐标系 II 变回 I。

这样,我们就得到了变换矩阵,是上面三个矩阵的级联。那么对于新图像中的每一点,就可以根据对应原图中的点,得到它的灰度。如果超出原图范围,则填成白色。要注意的是,由于有浮点运算,计算出来点的坐标可能不是整数,采用取整处理或插值法来处理。

图 12-6 所示是对图像旋转处理的流程图。

3.图像缩放

假设放大因子为 ratio,缩放的变换矩阵为:

$$\begin{bmatrix} x_0 \\ y_0 \\ 1 \end{bmatrix} = \begin{bmatrix} \dfrac{1}{\text{ratio}} & 0 & 0 \\ 0 & \dfrac{1}{\text{ratio}} & 0 \\ 0 & 0 & 1 \end{bmatrix} \begin{bmatrix} x_1 \\ y_1 \\ 1 \end{bmatrix} \tag{8}$$

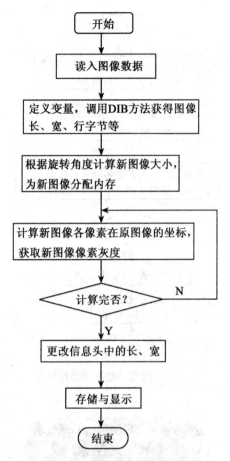

图 12-6　图像旋转处理的流程图

图 12-7 所示是图像缩放处理的流程图。

【实习步骤】

1.打开上次实习创建的工程。

2.为该工程添加一个新类,用于实现有关 BMP 格式图像文件处理的各种操作。

3.在新类中加入相应属性和操作,比如上次的格式转换、读写操作以及这次实习要实现的平移、缩放、旋转等操作。

4.如图 12-8 所示,在菜单中加入"平移/放大/缩小/旋转"菜单项,并为其添加相应消息处理函数。

5.在函数定义处添加自己的源代码(通过调用新建类的对象中相应的操作来实现)。

6.编译检查语法错误,若编译通过,运行程序,检查上述变换是否正确。

实习完毕后,提交一份实习报告。

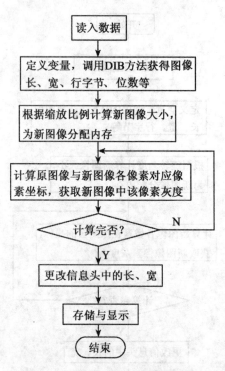

图 12-7 图像缩放处理的流程图

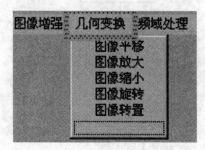

图 12-8 添加菜单项

【思考题】

1. 如果新图中有一点(x_1, y_1)，按照公式(2)得到的(x_0, y_0)不在原图中该怎么办？
2. 编程实现基于三种灰度内插法的图像放大技术，比较三种灰度内插方法的效果。
3. 实现图像旋转有哪些方法？你认为哪种方法最易实现？
4. 编程实现图像镜像、卷帘功能。

实习三 图像频域处理

【实习目的】

熟悉傅立叶变换算法和图像频域处理的基本流程,掌握频率域滤波的图像增强算法,开拓学生视野,提高图像增强处理技能。

【实习内容】

按照指导教师要求,借助正反傅立叶变换,编程实现对影像进行巴特沃斯、理想低通滤波和高通滤波处理。

【预备知识】

1. 熟悉傅立叶变换的基本原理。
2. 熟悉数字图像的快速傅立叶变换算法。
3. 熟悉 BMP 位图文件的读写和显示。

【实习原理】

1.傅立叶变换

一幅数字图像作为一个二维信号,同样可以进行傅立叶变换,从而进行频域分析。由于图像的数据量大,因此要求傅立叶变换能实现快速计算。

(1)离散傅立叶变换(DFT)。

如果 $f(n)$ 为一个长度为 N 的数字序列,则其离散傅立叶变换 $F(k)$ 为:

$$F(k) = \sum_{n=0}^{N-1} f(n) W_N^{kn} \quad 0 \leq k \leq N-1 \tag{1}$$

其中,
$$W_N = e^{-j\frac{2\pi}{M}} \tag{2}$$

本质上讲,DFT 的计算问题就是对给定序列 $f(n)$,按照式(1)计算长度为 N 的复值序列 $\{F(k)\}$。

类似地,离散傅立叶逆变换(IDFT)为:

$$f(n) = \frac{1}{N} \sum_{k=0}^{N-1} F(k) W_N^{-nk} \quad 0 \leq n \leq N-1 \tag{3}$$

由于 DFT 和 IDFT 基本上包含相同类似的计算,因此,关于 DFT 有效计算算法也适用于 IDFT 的计算。

由(3)式可以发现,对 k 的每一个取值,直接计算 $F(k)$ 涉及 N 次复数乘法(4N 次实数乘法)和 $N-1$ 次复数加法(4$N-2$ 次实数加法)。因此,计算 DFT 的所有 N 个值总共需要 N^2 次复数乘法和 N^2-N 次复数加法。

直接计算 DFT 效率低的主要原因是没有利用相位因子 W_N 的对称性和周期性。具体地说,这两个性质为:

对称性质:
$$W_N^{k+N/2} = -W_N^k \tag{4}$$

周期性质:
$$W_N^{k+N} = W_N^k \tag{5}$$

(2)DFT 的直接计算。

对 N 点复值序列 $f(n)$,它的 DFT 可表示为:

$$F_R(k) = \sum_{n=0}^{N-1}\left[f_R(n)\cos\frac{2\pi kn}{N} + f_I(n)\sin\frac{2\pi kn}{N}\right] \tag{6}$$

$$F_1(k) = -\sum_{n=0}^{N-1}\left[f_R(n)\sin\frac{2\pi kn}{N} - f_I(n)\cos\frac{2\pi kn}{N}\right] \tag{7}$$

直接计算式(6)和式(7)需要 $2N^2$ 次三角函数运算，$4N^2$ 次实数乘法运算，$4N(N-1)$ 次实数加法运算，大量的排序和寻址运算。

这些运算都是 DFT 计算的典型运算，第二项和第三项运算产生 DFT 的值 $F_R(k)$ 和 $F_I(k)$。对读取数据($0\leqslant n\leqslant N-1$)和相位因子并且保存结果来讲，排序和寻址运算是不可少的。可以按不同方式对这些计算过程进行优化，从而产生各式各样的 DFT 算法。

(3) DFT 的分解征服计算方法。

该方法的原理是，把计算长度为 N 的序列的离散傅立叶变换逐次地分解为计算长度较短的序列的离散傅立叶变换。该基本方法导出了一类统称为 FFT 算法的有效计算方法。

为了解释基本概念，现考虑 N 点 DFT 的计算，其中 N 可以分解成两个整数的积，即 $N=LM$。

这里并不严格限制 N 必须是素数，因为我们可以通过对这一序列补零来保证形式如式(4)的因式分解。

长度为 N 的序列 $f(n)$，$0\leqslant n\leqslant N-1$ 或者存储在下标为 n 的一维数组中，或者作为一个下标为 l 和 m 的二维数组存储，其中 $0\leqslant l\leqslant L-1$ 且 $0\leqslant m\leqslant M-1$。这里要注意 l 为行下标，m 为列下标。因此，在各种各样的方法中，将序列存储在二维数组中，它的每个数据都依赖于下标 n 到下标 (l,m) 的映射。

假定选取映射为：

$$n = Ml + m \tag{8}$$

该映射将产生一个排列，其第一行由 $f(n)$ 的前 M 个元素组成，第二行由 $f(n)$ 的紧接着的 M 个元素构成，依此类推。另一方面，若映射为：

$$n = l + mL \tag{9}$$

则 $f(n)$ 的前 L 个元素存储在第一列，紧接着的 L 个元素存储在第二列，依此类推。

计算所得的 DFT 值可以存储在一个类似的排列中，具体地说，映射是从下标 k 到下标对 (p,q) 的映射，其中 $0\leqslant p\leqslant L-1$，$0\leqslant q\leqslant M-1$，假如我们选取映射为：

$$k = Mp + q \tag{10}$$

则 DFT 是以行向为基础存储的，其中第一行存储的是 $F(k)$ 的前 M 个元素，第二行存储紧接着的 M 个元素，依此类推。若选取另一种映射：

$$k = qL + p \tag{11}$$

则 $F(k)$ 为列向存储，其中前 L 个元素存储在第一列，紧接着的 L 个元素存储在第二列，依此类推。

现在假定 $f(n)$ 映射到二维数组 $x(l,m)$，将 $F(k)$ 映射到相位因子之后的二维数组元素双重求和。具体地说，让我们对 $F(p,q)$ 采用式(9)定义的列向映射，同时，对 DFT 采用(10)式定义的行向映射，因此有：

$$F(p,q) = \sum_{m=0}^{M-1}\sum_{l=0}^{L-1} f(l,m) W_N^{(Mp+q)(mL+l)} \tag{12}$$

但是

$$W_N^{(Mp+q)(mL+l)} = W_N^{MLmp} W_N^{mLq} W_N^{Mpl} W_N^{lq} \tag{13}$$

然而，$W_N^{Nmp}=1$，$W_N^{mqL}=W_{N/L}^{mq}=W_M^{mq}$，而且 $W_N^{Mpl}=W_{N/M}^{pl}=W_L^{pl}$。

利用这些简化，式(12)可表示为：

$$F(p,q) = \sum_{l=0}^{L-1} \left\{ W_N^{lq} \left[\sum_{m=0}^{M-1} f(l,m) W_M^{mq} \right] \right\} W_L^{lp} \tag{14}$$

在式(14)中的表达式涉及计算长度为 M 和 L 的 DFT。为了详细阐述，我们将计算细分为三步：

①对第一行 $l=0,1,\cdots,L-1$，首先计算 M 点 DFT：

$$S(l,q) \equiv \sum_{m=0}^{M-1} f(l,m) W_M^{mq} \quad 0 \leq q \leq M-1 \tag{15}$$

②计算一个新的二维数组 $T(l,q)$，该二维数组定义为：

$$T(l,q) = W_N^{lq} F(l,q) \quad 0 \leq l \leq L-1, 0 \leq q \leq M-1 \tag{16}$$

③对 $q=0,1,\cdots,M-1$ 时二维数组的每一列，计算 L 点 DFT：

$$F(p,q) \equiv \sum_{l=0}^{L-1} T(l,q) W_L^{lp} \tag{17}$$

式(14)的计算量包括：第一步包含 L 个 M 点 DFT 的计算，因此第一步需要 L 次复数乘法和 $LM(M-1)$ 次复数加法；第二步需要 LM 次复数乘法；第三步需要 ML^2 次复数乘法和 $ML(L-1)$ 次复数加法。所以，计算量为复数乘法 $N(M+L+1)$ 次，复数加法 $N(M+L-2)$ 次（其中 $N=ML$）。于是，乘法次数从 N^2 次减少到 $N(M+L+1)$ 次，加法次数从 $N(N-1)$ 次减为 $N(M+L-2)$ 次。

当 N 为一个高度组合数时，也就是说，可将 N 因子分解为若干素数的积

$$N = r_1 r_2 \cdots r_v \tag{18}$$

那么，上面的 DFT 分解还可重复 $v-1$ 次。该过程导致了更小的 DFT，产生了一个更加有效的计算方法。

在功效方面，第一次将序列 $f(n)$ 分解为每列由 L 个元素构成的 M 列二维数组，产生了长度为 L 和 M 的 DFT，然后对每行（或每列）进一步有效分解成更小的二维数组将产生更小的 DFT，继续该分解过程，直到 N 被完全因子分解为素因子时为止。

如果令 $W = e^{j\frac{2\pi}{N}}$，那么(3)式和(4)式变为：

$$F(\mu) = \sum_{x=0}^{N-1} f(x) e^{-j\frac{2\pi ux}{N}} = \sum_{x=0}^{N-1} f(x) W^{-ux} \tag{19}$$

$$f(x) = \frac{1}{N} \sum_{\mu=0}^{N-1} F(\mu) e^{j\frac{2\pi ux}{N}} = \frac{1}{N} \sum_{\mu=0}^{N-1} F(u) W^{ux} \tag{20}$$

写成矩阵形式为

$$\begin{bmatrix} F(0) \\ F(1) \\ \vdots \\ F(N-1) \end{bmatrix} = \begin{bmatrix} W^0 & W^0 & W^0 & \cdots & W^0 \\ W^0 & W^{1\times 1} & W^{2\times 1} & \cdots & W^{(N-1)\times 1} \\ \vdots & \vdots & \vdots & & \vdots \\ W^0 & W^{1\times(N-1)} & W^{2\times(N-1)} & \cdots & W^{(N-1)\times(N-1)} \end{bmatrix} \begin{bmatrix} f(0) \\ f(1) \\ \vdots \\ f(N-1) \end{bmatrix} \tag{21}$$

$$\begin{bmatrix} f(0) \\ f(1) \\ \vdots \\ f(N-1) \end{bmatrix} = \frac{1}{N} \begin{bmatrix} W^0 & W^0 & W^0 & \cdots & W^0 \\ W^0 & W^{-1\times 1} & W^{-2\times 1} & \cdots & W^{-(N-1)\times 1} \\ \vdots & \vdots & \vdots & & \vdots \\ W^0 & W^{-1\times(N-1)} & W^{-2\times(N-1)} & \cdots & W^{-(N-1)\times(N-1)} \end{bmatrix} \begin{bmatrix} F(0) \\ F(1) \\ \vdots \\ F(N-1) \end{bmatrix} \tag{22}$$

同理,二维离散函数 $f(x,y)$ 的傅立叶变换为

$$F(u,v) = \sum_{x=0}^{M-1} \sum_{y=0}^{N-1} f(x,y) \mathrm{e}^{-j2\pi(\frac{ux}{M}+\frac{vy}{N})} \tag{23}$$

傅立叶逆变换为:

$$f(x,y) = \frac{1}{MN} \sum_{x=0}^{M-1} \sum_{y=0}^{N-1} F(u,v) \mathrm{e}^{j2\pi(\frac{ux}{M}+\frac{vy}{N})} \tag{24}$$

其中, $x=0,1,2,\cdots,M-1; y=0,1,2,\cdots,N-1$。

在数字图像处理中,图像一般为方阵,即 $M=N$,则二维离散傅立叶变换公式为:

$$F(u,v) = \sum_{x=0}^{N-1} \sum_{y=0}^{N-1} f(x,y) \mathrm{e}^{-j2\pi \frac{ux+vy}{N}} \tag{25}$$

$$f(x,y) = \frac{1}{N^2} \sum_{x=0}^{N-1} \sum_{y=0}^{N-1} F(u,v) \mathrm{e}^{j2\pi \frac{ux+vy}{N}} \tag{26}$$

图 12-9 所示是图像快速离散傅立叶变换的流程图。

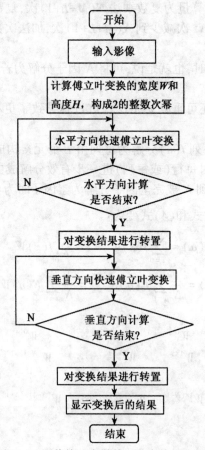

图 12-9 图像快速离散傅立叶变换的流程图

2. 频率域图像增强

假定原图像为 $f(x,y)$，经傅立叶变换为 $F(u,v)$，频率域增强就是选择合适的滤波器 $H(u,v)$ 对 $F(u,v)$ 的频谱成分进行调整，然后经逆傅立叶变换得到增强的图像 $g(x,y)$。图12-10所示是频率域增强的一般过程。

图 12-11 所示是巴特沃斯低通滤波的流程图。

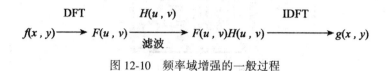

图 12-10　频率域增强的一般过程

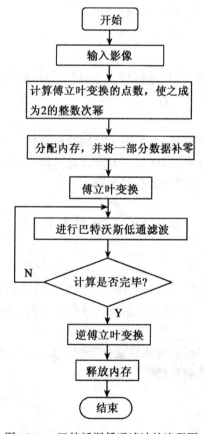

图 12-11　巴特沃斯低通滤波的流程图

【实习步骤】

1. 打开上次实习所用的项目。
2. 在自己创建的类中添加傅立叶正变换、傅立叶逆变换和低通滤波操作。

3. 如图12-12所示添加菜单项"傅立叶变换"、"巴特沃斯低通滤波"和"傅立叶逆变换"。

图12-12 添加菜单项

4. 为新加的菜单项分别建立消息处理函数。

5. 在相应函数体内分别添加实现傅立叶正、反变换和巴特沃斯低通、高通滤波的源代码（通过类对象中的操作来实现）。

6. 检查语法错误。若编译通过，运行程序，观察与分析傅立叶变换、低通滤波的结果。
实习完毕后，提交一份实习报告。

【思考题】
1. 图像的傅立叶频谱图有何特点？用于分析图像有什么优势？
2. 编程实现高斯滤波器的频率域处理方法，同巴特沃斯滤波器的处理结果进行比较。
3. 频率域滤波与空间域平滑效果有何区别？

实习四 伪彩色增强

【实习目的】
熟悉灰度图像的伪彩色增强原理和方法，编程实现图像伪彩色增强变换，掌握伪彩色增强技术及其应用，提高学生彩色图像处理技能、分析能力和实际动手能力。

【实习内容】
在新建的类中加入灰度分割、空间域变换彩色合成、频率域变换彩色合成操作，按照指导老师要求，编程实现这些伪彩色增强的功能。

【预备知识】
1. 熟悉图像的读写和显示。
2. 熟悉图像傅立叶变换算法。

【实习原理】
1. 灰度分割

一幅灰度图像可看做一个2-D的灰度函数，用一个平行于图像坐标平面去分截图像灰度函数，从而把灰度函数分成两个灰度值区间。图12-13给出一个分割的剖面示意图（横轴为坐标轴，纵轴为灰度值轴）。

根据图12-14，对每一像素输入灰度值，如果它在灰度值l_m之下就赋予一种颜色，否则

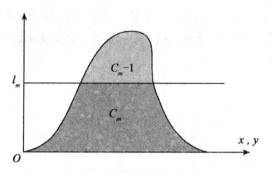

图 12-13 灰度切割示意图

赋予另一种颜色。通过这种变换,原来的多灰度值图变成了一幅只有两种颜色的图。上下平移切割平面就可得到不同的结果。

这种方法可推广如下:设在灰度级 $l_1, l_2, \cdots, l_M$ 处定义了 M 个平面,让 l_0 代表黑($f(x,y)=0$),L_M 代表白($f(x,y)=L$),在 $0<l_m<L$ 的条件下,M 个平面将把图像灰度值分成 $M+1$ 个区间,对同一灰度值区间内的像素赋相同颜色,不同灰度值区间赋予不同颜色,这就是灰度分割方法。即

$$f(x,y) = C_m \quad f(x,y) \in R_m, \ m = 0, 1, \cdots, M \tag{1}$$

其中,R_m 为由分割平面限定的灰度区间,而 C_m 是所赋的颜色。图 12-14 是密度分割法的流程图。

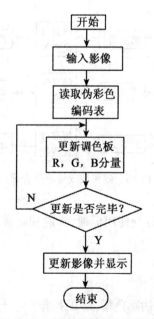

图 12-14 密度分割法的流程图

2. 空间域变换彩色合成

若对原始图像中用三个独立的变换来处理,如图 12-15 所示,那么三个变换的结果分别同时输入彩色电视屏幕的三个电子枪,这样就可得到三个变换函数变换合成的彩色图像。

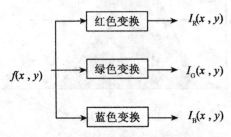

图 12-15　空间域变换彩色合成

方法的关键之处在于变换函数的确定。可以使用光滑的、非线性的变换函数,实际中变换函数常用取绝对值的正弦函数。灰度分割方法可看做是用一个分段线性函数实现从灰度到彩色的变换,可看做本方法的一个特例。

3. 频率域变换彩色合成

伪彩色增强也可通过频率域处理来实现,如图 12-16 所示。输入图像经傅立叶变换,通过三个不同的滤波器得到不同的频率分量,然后对各频率分量分别进行傅立叶逆变换,其结果经进一步处理(如直方图均衡化或规定化),最后进行彩色合成就能得到增强后的彩色图像。

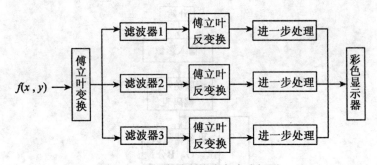

图 12-16　频率域变换彩色合成框图

为得到不同的频率分量,可分别使用低通、带通和高通滤波器作为图 12-16 中的 3 个滤波器。

【实习步骤】
1. 打开上次实习所用的项目。
2. 在自己创建的类中添加"伪彩色增强"的操作。
3. 添加菜单项"灰度分割","空间域变换彩色合成","频率域变换彩色合成"。
4. 为新加的菜单项建立消息处理函数。

5. 在相应函数体内添加上述三种方法伪彩色增强的源代码(通过类对象中的操作来实现)。

6. 编译检查语法错误。若编译通过,运行程序,观察与分析伪彩色增强效果。

实习完毕后,提交一份实习报告。

【思考题】

1. 如何合理地选择分割平面对灰度影像进行灰度分割?

2. 试用多种函数通过从灰度到彩色的变换方法对灰度影像进行伪彩色增强,并比较各种函数增强效果的差异。

3. 频率域滤波方法中可以采用多种滤波器以获得不同的频率分量,试比较各不同频率分量处理灰度影像的效果差异。

实习五 基于高通滤波的影像融合

【实习目的】

在熟悉像素级融合理论的基础上,应用所学知识,实现基于高通滤波的影像融合方法,以获得分辨率和信息量提高的融合影像,提高学生图像处理与分析能力和实际动手能力。

【实习内容】

首先设计低通滤波器,对给定的高分辨率全色影像进行处理,提取影像的低频信息和高频信息;其次把高分辨率影像的高频信息分别加到与高分辨率全色影像配准的低分辨率多光谱影像上,通过彩色合成得到融合的彩色影像。以上各功能,按照指导老师要求,用VC++编程实现,并分析融合方法的特点和评价融合影像质量。

【预备知识】

1.熟悉 BMP 位图文件的读写和显示。

2.熟悉基于像素的影像融合技术。

【实习原理】

提高多光谱影像空间分辨率的方法之一是将较高空间分辨率影像的高频信息(细节和边缘等)逐像素叠加到低空间分辨率的多光谱影像上。

根据对高频信息分量的叠加方式,可分为不加权融合法和加权融合法两种。

1. 不加权融合法

不加权融合法是采用一个较小的空间高通滤波器对高空间分辨率影像滤波,直接将高通滤波得到的高频成分依像素加到各低分辨率多光谱影像上,获得空间分辨率增强的多光谱影像。常称为空间高通滤波融合法。融合表达式如下:

$$F_k(i,j) = M_k(i,j) + \text{HPH}(i,j) \quad (1)$$

式中,$F_k(i,j)$ 表示第 k 波段像素(i,j) 的融合值;$M_k(i,j)$ 表示低分辨率多光谱影像第 k 波段像素(i,j)的值;高分辨率影像 $P(i,j)$ 经过低通滤波后分解成 $\text{LPH}(i,j)$ 和 $\text{HPH}(i,j)$ 两部分,即

$$P(i,j) = \text{LPH}(i,j) + \text{HPH}(i,j) \quad (2)$$

图 12-17 所示是该融合方法的流程图。该方法将较高空间分辨率影像的高频信息与多

光谱影像的光谱信息融合,获得空间分辨率增强的多光谱影像,并具有一定的去噪功能。问题的关键是设计滤波器。合理的办法是用低通滤波器去匹配多光谱影像的点扩散函数,但精确测定多光谱影像的点扩散函数是很困难的,而且与点扩散函数匹配的低通滤波器受瞬时视场、传感器类型和重采样函数的影响。因此通过自己已学知识,在实习过程中探讨何种滤波器比较适用。

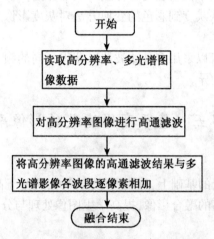

图 12-17 空间高通滤波融合方法的流程图

2. 加权融合法

为了改善融合效果,提出了高频调制融合算法。它是将高分辨率影像 $P(i,j)$ 与空间配准的低分辨率第 k 波段多光谱影像 $M_k(i,j)$ 进行相乘,并用高分辨率影像 $P(i,j)$ 经过低通滤波后得到的影像 $\mathrm{LPH}(i,j)$ 进行归一化处理,得到增强后的第 k 波段融合影像。其公式为:

$$F_k(i,j) = M_k(i,j) \cdot P(i,j)/\mathrm{LPH}(i,j) \tag{3}$$

将上式代入式(2)得

$$F_k(i,j) = M_k(i,j) + M_k(i,j) \cdot \mathrm{HPH}(i,j)/\mathrm{LPH}(i,j) \tag{4}$$

令 $K(i,j) = M_k(i,j)/\mathrm{LPH}(i,j)$,则有

$$F_k(i,j) = M_k(i,j) + K(i,j) \cdot \mathrm{HPH}(i,j) \tag{5}$$

可见式(5)对高分辨率影像高频部分 $\mathrm{HPH}(i,j)$ 以权 $K(i,j)$ 调整,然后加到多光谱影像上,因此称为加权融合法。图 12-18 所示是该方法的流程图。该方法技术关键在于设计合适的低通滤波器。

以上是对两种空间域融合方法的介绍,在本次实习中,要求使用两种方法对给定影像进行融合。

【实习步骤】

1. 打开上次实习所建的项目。
2. 添加菜单项"影像融合"。
3. 为"影像融合"添加子菜单不加权融合和加权融合。
4. 为新加的菜单项"不加权融合"和"加权融合"建立消息响应函数。

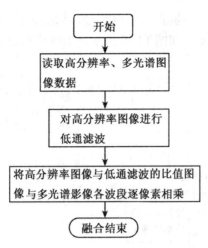

图 12-18 高频调制融合方法的流程图

5. 在函数体内添加实现"不加权融合"和"加权融合"的源代码。

6. 编译检查语法错误,若编译通过,运行程序。显示融合图像,检查是否正确实现融合算法。

实习完毕后,提交一份实习报告。

【思考题】

1. 采用不同大小的低通滤波器进行"加权融合",分析融合效果的变化规律。
2. 如何评价融合影像的质量?

实习六 基于 HIS 变换的影像融合方法

【实习目的】

熟悉彩色图像的读写,掌握彩色变换及其用于影像融合的方法,评价融合影像质量,提高学生图像处理与分析能力和实际动手能力。

【实习内容】

1. 编程实现彩色图像的 HIS 正、反变换算法。
2. 编程实现 HIS 变换影像融合方法,对融合影像质量评价,分析融合方法的特点。

【预备知识】

1. 熟悉彩色变换原理。
2. 熟悉 BMP 位图文件的读写和显示。

【实习原理】

1. 彩色变换

颜色可以用 R、G、B 三分量来表示,也可以用亮度(I)、色别(H)和饱和度(S)来表示,它们称为颜色的三要素。把彩色的 R、G、B 变换成 I、H、S 称为 HIS 正变换,而由 I、H、S 变换成 R、G、B 称为 HIS 反变换。

2. HIS 变换融合方法

影像融合是采用某种算法将覆盖同一地区(或对象)的两幅或多幅空间配准的影像生成满足某种要求的影像的技术。如图 12-19 所示是 HIS 变换融合的流程。

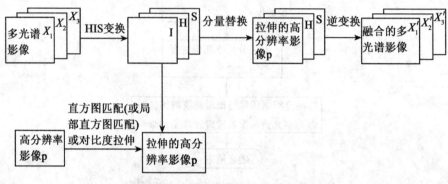

图 12-19　HIS 变换融合的流程图

融合步骤如下：

(1) 将空间分辨率低的 3 个波段多光谱影像变换到 HIS 空间,得到色别 H、明度 I、饱和度 S 三分量。

(2) 将高空间分辨率影像进行直方图匹配(直方图规定化)或对比度拉伸,使之与 I 分量有相同的均值和方差。

(3) 用拉伸后的高空间分辨率影像代替 I 分量,同 H、S 分量进行 HIS 逆变换得到空间分辨率提高的融合影像。

【实习步骤】

1. 打开上次实习所用的项目。
2. 在自己创建的类中添加 HIS 变换操作。
3. 添加菜单项"HIS 正变换"、"HIS 反变换"以及"HIS 融合"。
4. 为新加的菜单项建立消息处理函数。
5. 在相应函数体内添加实现 HIS 正、反彩色变换以及 HIS 融合的源代码(通过类对象中的操作来实现)。
6. 编译检查语法错误。若编译通过,运行程序,观察彩色变换以及融合后的图像,分析它们与原影像的差异。

实习完毕后,提交一份实习报告。

【思考题】

1. HIS 变换融合法不足之处有哪些？应如何改进？
2. 主分量变换融合法是一种常用的图像融合方法,相对于 HIS 变换而言,其有何优点？
3. HIS 变换可与小波变换、金字塔等多分辨率分析技术结合形成新的融合方法,试分析这些方法的融合效果。

实习七　灰度图像边缘检测

【实习目的】

依据边缘检测理论,实现灰度图像一阶和二阶边缘检测方法,启发学生依据边缘特征进行图像分析与识别,提高学生图像处理与分析能力和实际动手能力。

【实习内容】

1. 编程实现一阶差分边缘检测算法,包括 Roberts 梯度算子、Prewitt 算子和 Sobel 算子。
2. 编程实现二阶差分 Laplace 边缘检测算法。
3. 分析与比较各种边缘检测算法的性能。

【预备知识】

1. 熟悉 BMP 位图的读写和显示。
2. 熟悉图像分割理论。

【实习原理】

对于阶跃状边缘,在边缘点处一阶导数有极值,因此可以利用这一特性通过计算每个像素的梯度来检测边缘点。对于离散图像来说,常用一阶差分近似表示一阶导数,即

$$f'_x = f(x+1,y) - f(x,y) \tag{1}$$

$$f'_y = f(x,y+1) - f(x,y) \tag{2}$$

为简化梯度的计算,常用下面的近似表达式:

$$\mathrm{grad}(x,y) = \max(|f'_x|, |f'_y|),\text{或者 } \mathrm{grad}(x,y) = |f'_x| + |f'_y| \tag{3}$$

常用的梯度算子有 Roberts 梯度算子、Prewitt 算子和 Sobel 算子,图 12-20 所示为它们对应的模板。

对于阶跃状边缘,其二阶导数在边缘点出现零交叉,且边缘点两旁二阶导数取异号。对数字图像的每个像素计算关于 x 和 y 方向的二阶偏导数之和 $\nabla^2 f(x,y)$。

$$\nabla^2 f(x,y) = f(x+1,y) + f(x-1,y) + f(x,y+1) + f(x,y-1) - 4f(x,y) \tag{4}$$

式(4)就是著名的 Laplace 算子,对应的模板如图 12-20 所示。

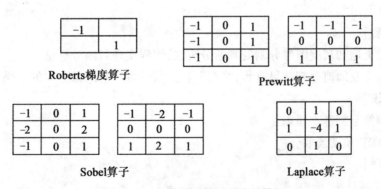

图 12-20　边缘检测算子对应的模板

【实习步骤】
1. 打开上次实习所用的项目。
2. 在自己创建的类中添加各种边缘检测算子操作。
3. 如图 12-21 所示添加菜单项"边缘检测",再为"边缘检测"菜单添加子菜单项"Roberts 边缘检测","Prewitt 边缘检测","Sobel 边缘检测"菜单项。

图 12-21　添加菜单项

4. 为新加的菜单项建立消息处理函数。
5. 在相应函数体内添加实现相应操作的源代码(通过类对象中的操作来实现)。
6. 编译检查语法错误,若编译通过,运行程序,观察与分析边缘检测算子检测结果。
实习完毕后,提交一份实习报告。
【思考题】
1. 利用一阶导数和二阶导数特性检测图像边缘,分析检测结果的主要区别。
2. 在分析几种经典的边缘检测方法的基础上,设计一种通用的边缘特征检测方法,使之具有 Roberts、Laplace、Sobel、Prewitt 等检测算子的功能。

实习八　图像二值化:判断分析法

【实习目的】
加深对图像分割的理解,掌握最简单图像分割的原理与方法,提高学生图像处理与分析能力和实际动手能力。
【实习内容】
1. 给自建类增加灰度图像分割的判断分析法,编程实现判断分析法。
2. 与人工交互阈值分割结果进行比较,分析判断分析法对图像分割的效果。
【预备知识】
1. 熟悉图像分割基本原理。
2. 熟悉 BMP 位图读取。
【实习原理】
1. 状态法(峰谷法)
如果一幅灰度图像的直方图有双峰和明显的谷,如图 12-22 所示,那么选择两峰之间的谷所对应的灰度 T 作为阈值,按下式进行二值化,即可将目标从图像中分割出来:

$$g(x,y) = \begin{cases} 0, f(x,y) < T \\ 1, f(x,y) \geq T \end{cases} \qquad (1)$$

其中,$f(x,y)$ 为原图像;T 为阈值;$g(x,y)$ 为分割后的图像。

这种方法比较简单,但不适合于两峰值相差比较大、又宽且平的谷底的图像。

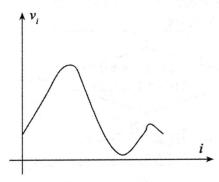

图 12-22　灰度直方图

2. 判断分析法

假定简单图像 $f(x,y)$ 的灰度区间为 $[0,L-1]$,则选择一阈值 T 将图像的像素分为 c_1、c_2 两组。

$$\begin{cases} c_1, f(x,y) < T, \quad 像素数为 w_1,灰度平均值为 m_1,方差为 \sigma_1^2 \\ c_2, f(x,y) \geq T, \quad 像素数为 w_2,灰度平均值为 m_2,方差为 \sigma_2^2 \end{cases} \qquad (2)$$

图像总像素数为 w_1+w_2,灰度均值为 $m=(m_1 w_1+m_2 w_2)/(w_1+w_2)$,组内方差为 $\sigma_w^2 = w_1 \sigma_1^2 + w_2 \sigma_2^2$,组间方差为 $\sigma_B^2 = w_1(m_1-m)^2 + w_2(m_2-m)^2 = w_1 w_2(m_1-m_2)^2$。

显然,组内方差越小,则组内像素越相似;组间方差越大,则组间的差别越大。因此,σ_B^2/σ_w^2 为最大值所对应的 T,就是所求判断分析法的分割阈值。图 12-23 是判断分析法的流程图。

【实习步骤】

1. 打开上次实习所用的项目。
2. 在自己创建的类中添加各种图像分割算法操作。
3. 如图 12-24 所示添加菜单项"判断分析法"。
4. 为新加的菜单项建立消息处理函数。
5. 在相应函数体内添加实现相应操作的源代码(通过类对象中的操作来实现)。
6. 编译检查语法错误。若编译通过,运行程序,观察判断分析法分割效果,对结果进行评价。

实习完毕后,提交一份实习报告。

【思考题】

1. 比较所实现的几种分割算法的效果,分析其特点。

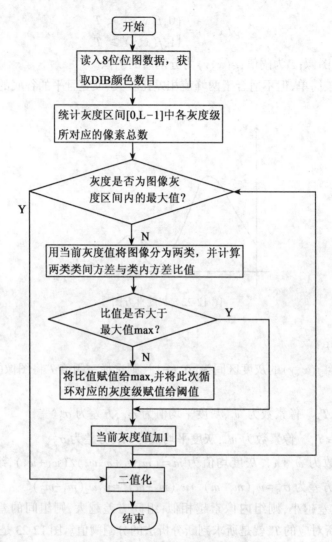

图 12-23　判断分析法流程图

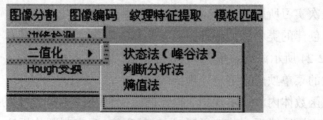

图 12-24　添加菜单项

2. 试编程实现一种基于经验知识的改进 Otsu 算法(江苏大学学报:自然科学版,2005,26(5):401-404),验证该算法具有能缩小搜索阈值范围,减少计算类间方差的次数的优点。

实习九　Hough 变换检测直线

【实习目的】
进一步掌握 Hough 变换检测直线的原理,编程实现 Hough 变换提取直线的算法,分析 Hough 变换检测性能,提高学生图像处理与分析能力和实际动手能力。

【实习内容】
给自建类增加 Hough 变换提取直线的功能,编程实现二值图像直线的提取功能。

【预备知识】
1. 熟悉 Hough 变换的基本原理。
2. 熟悉位图读取。
3. 熟悉 Hough 变换提取直线的方法。

【实习原理】
Hough 变换是图像处理中检测几何形状的基本方法之一,应用很广泛。Hough 变换最基本应用是从二值图像中检测直线段。

如图 12-25 所示,在 x,y 平面上,任何一条直线 $y=kx+b$(其中 k,b 分别表示斜率和截距),若 θ 是该直线与 x 轴的夹角,ρ 是原点到直线的距离,则这条直线可表示为:

$$\rho = x\cos\theta + y\sin\theta \qquad (1)$$

Hough 变换就是一种直线到点的变换。即 xy 平面的任意一条直线,在 ρ 和 θ 定义的二维空间上对应一个点。

对于 x,y 平面的一个特定点 (x_0,y_0),通过该点的直线有很多条,每一条都对应 ρ,θ 平面中的一个点。并且与 x,y 平面中所有这些直线对应的点在参数空间中的轨迹是一条正弦曲线,因而过 x,y 平面上的任意一点的所有直线对应于 ρ,θ 平面的一条正弦曲线。

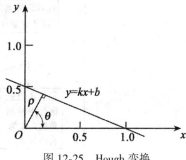

图 12-25　Hough 变换

如果一组边缘点位于由参数 ρ_0 和 θ_0 决定的直线上,则每个边缘点对应了 ρ,θ 平面的一条正弦曲线,并且所有这些正弦曲线必相交于点 (ρ_0,θ_0),这就是 Hough 变换检测直线的原理。

Hough 变换检测直线的算法:

(1)初始化一个 ρ,θ 平面的数组。一般 ρ 方向上的量化数目为对角线方向像素数,θ 方向上的量化间距为 2°。

(2)顺序搜索图像的所有黑点,对每一个黑点,按式(1)计算 ρ,θ 取不同的值,分别将对应的数组元素加 1。

(3)求出数组中的最大值并记录对应的 ρ,θ。

(4)绘出 ρ,θ 对应的直线。如图 12-26 所示。

图 12-27 所示是利用 Hough 变换提取一条直线的流程图。

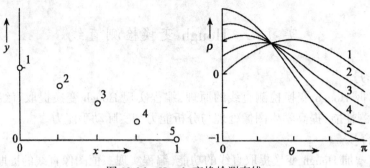

图 12-26　Hough 变换检测直线

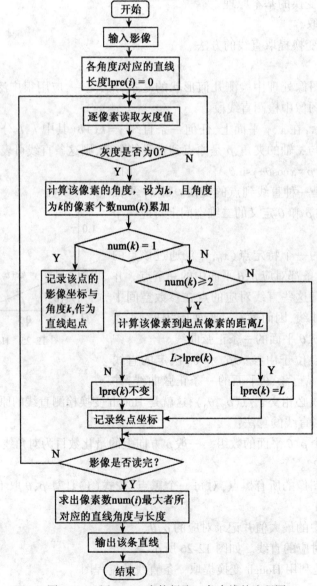

图 12-27　用 Hough 变换提取一条直线的流程图

【实习步骤】
1. 打开上次实习所用的项目。
2. 在自己创建的类中添加 Hough 变换提取直线算法操作。
3. 如图 12-28 所示添加菜单项"Hough 变换"。

图 12-28　添加菜单项

4. 为新加的菜单项建立消息处理函数。
5. 在相应函数体内添加实现相应操作的源代码(通过类对象中的操作来实现)。
6. 编译检查语法错误,若编译通过,运行程序,观察 Hough 变换检测直线的情况。
实习完毕后,提交一份实习报告。

【思考题】
1. 如何量化 ρ 和 θ？其量化粗细对检测有何影响？
2. 如何采用 Hough 变换检测多条直线？

实习十　霍夫曼编码

【实习目的】
加深对图像编码的理解与应用,掌握霍夫曼编码方法,提高学生图像处理与分析能力和实际动手能力。

【实习内容】
编程实现霍夫曼编码,并计算图像熵、平均码字长度及编码效率。

【预备知识】
1. 熟悉位图图像读取。
2. 熟悉 VC 编程。

【实习原理】
霍夫曼编码的基本思想是对图像中出现频率大的灰度级用较短的代码表示,出现频率小的灰度级用较长的代码表示,即用变长码表示图像数据,达到压缩的目的。

在计算霍夫曼编码表时需要对原始图像数据扫描两遍:第一遍扫描要精确地统计出原始图像中每个灰度级出现的概率;第二遍是建立霍夫曼树并进行编码。由于需要建立二叉树并遍历二叉树生成编码,数据压缩和还原速度很慢,但该编码方法简单有效,而且编码效率高,因而应用广泛。

霍夫曼编码的具体算法如下:

(1) 首先统计出每个灰度级出现的频率。
(2) 从左到右把上述频率按从大到小的顺序排列。
(3) 选出频率最小的两个灰度级频率,作为二叉树的两个叶子节点,将其和作为它们的根节点,两个叶子节点不再参与排序,新的根节点同其余灰度级出现的频率排序。
(4) 重复(3),直到最后得到和为 1 的根节点。
(5) 将形成的二叉树的子节点左边标 0,右边标 1。把最上面的根节点到最下面的叶子节点途中遇到的 0,1 序列串起来,就得到了各个灰度级的编码。
(6) 最后按照灰度级编码表对图像编码。

图 12-29 所示是霍夫曼编码的流程图。

【实习步骤】
1. 打开上次实习所用的项目。
2. 在自己创建的类中添加计算霍夫曼编码操作。
3. 如图 12-30 所示添加菜单项"霍夫曼编码"。

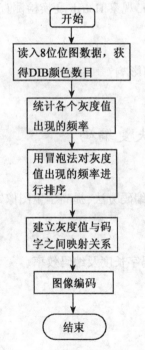

图 12-29　霍夫曼编码的流程图　　图 12-30　添加菜单项

4. 设计一个对话框,用于显示得到的霍夫曼编码及其他结果。
5. 为新加的菜单项建立消息处理函数。
6. 在相应函数体内添加实现霍夫曼编码的源代码(通过类对象中的操作来实现)。
7. 编译检查语法错误。若编译通过,运行程序,检查霍夫曼编码结果是否正确。
实习完毕后,提交一份实习报告。

【思考题】
编写一种优化的 Huffman 编码算法,并通过试验对其优越性加以验证。

实习十一 图像的行程编码

【实习目的】
加深对图像编码的理解与应用,掌握行程编码方法,提高学生图像处理与分析能力和实际动手能力。

【实习内容】
编程实现行程编码,并将 BMP 位图存储为 PCX 编码文件。

【预备知识】
1. 熟悉位图图像读取。
2. 熟悉 VC 编程。

【实习原理】
行程编码的基本原理是将一行中颜色值相同的相邻像素用一个计数值和该颜色值来代替。比如,aaabbcccccdddeeee 可以表示为 3a2b5c3d4e。如果一幅图像由很多块颜色相同的大面积区域组成,则采用行程编码可大大提高压缩效率。但也存在缺点,若图像中每两个相邻像素的颜色都不相同,则采用这种方法不但不能实现数据压缩,反而使数据量增加一倍。

PCX 文件是采用行程编码的例子之一。PCX 文件结构简单,由文件头和图像压缩数据两部分组成(如果是 256 色 PCX 图像文件,则还有一个 256 色调色板存于文件尾部)。

1.PCX 文件头(占 128 字节)数据结构

typedef struct {

  BYTE bManufacturer;//PCX 文件的标识,必须为 0x0A,用于判断一幅图像是否为 PCX 格式

  BYTE bVersion;//用于指明当前 PCX 文件的版本号

  BYTE bEncoding;//目前固定值为 1,表示采用行程编码

  BYTE bBpp;//指明每个像素需要的位数

  WORD wLeft;//指明图像相对于屏幕的左上角 x 坐标(以像素为单位)

  WORD wTop;//指明图像相对于屏幕的左上角 y 坐标(以像素为单位)

  WORD wRight;// 指明图像相对于屏幕的右下角 x 坐标(以像素为单位)

  WORD wBottom;// 指明图像相对于屏幕的右下角 y 坐标(以像素为单位)

  WORD wXResolution;//指明图像的水平分辨率(每英寸有多少个像素)

  WORD wYResolution;// 指明图像的垂直分辨率(每英寸有多少个像素)

  BYTE bPalette[48];//指明调色板数据。该域长度为 48 字节,只能保存 16 种颜色,256 色图像调色板保存在图像的尾部。此时 bPalette 域是没有任何意义的

  BYTE bReserved;//保留域,设定为 0

BYTE bPlanes;//指明图像色彩平面数目。该域和 bBpp 域决定图像的颜色总数
WORD wLineBytes;//指定图像的宽度(以字节为单位),它必须为偶数
WORD wPaletteType;//指定图像调色板的类型,1 表示彩色或者单色图像,2 表示图像是灰度图
WORD wSrcWidth;//指定制作该图像的屏幕宽度(像素为单位,0 为基准,即取值为屏幕宽度减 1)
WORD wSrcDepth;//指定制作该图像的屏幕宽度(像素为单位,0 为基准,即取值为屏幕高度减 1)
BYTE bFiller[54];//保留域,取值为 0
}PCXHEADER;

2.图像压缩数据

图像压缩数据紧跟在文件头后面。图像数据存储和图像颜色数目密切相关。这里介绍 256 色 PCX 文件。256 色 PCX 文件每个像素一个字节,编码时按照从左到右、从上到下的顺序进行(如果图像的宽度为奇数,那么每行需要添加一个字节)。进行编码时,是以字节为单位,一行一行地进行。首先计算原始数据中各个数据出现的次数,然后用该数据重复次数加上数据本身来代替原始数据。编码原则如下:

(1) 图像数据是以字节为单位进行编码的。

(2) 对于连续重复的像素值,统计其连续出现的次数 iCount(最大取值为 63),先存入长度信息(iCount|0xC0),然后再存入像素值。如果连续次数超过 63 次,则必须分多次处理。

(3) 如果遇到不重复的像素值,如果该像素值小于或等于 0xC0,则直接存入该像素值。否则首先存入一个 0xC1,然后再存入该像素值。这样做是为了避免像素值被误认为是数据长度。

3.256 色调色板

对于 256 色 PCX 文件,在图像数据后面还有一个长度为 769 字节的 256 色调色板。其中它的第一个字节为调色板标志字节,取值恒为 0x0C。接下来的 768(256×3)个字节为调色板的内容。

图 12-31 给出了行程编码的流程图。

【实习步骤】

1.打开上次实习所用的项目。

2.在自己创建的类中添加将文件按 PCX 行程编码文件进行编码的操作。

3.如图 12-32 所示添加菜单项"行程编码"。

4.为新加的菜单项建立消息处理函数。

5.在相应函数体内添加实现行程编码操作的源代码。

6.编译检查语法错误,若编译通过,运行程序,观察重编码得到的 PCX 文件,可以用其他图像软件打开进行结果检查。

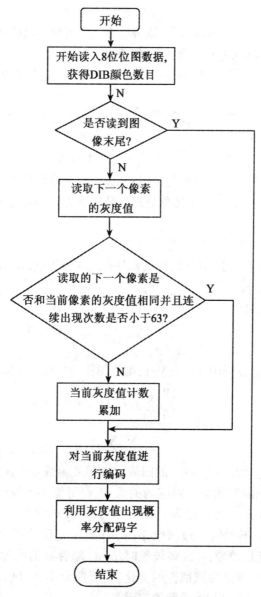

图 12-31　行程编码的流程图

图 12-32　添加菜单项

实习完毕后,提交一份实习报告。
【思考题】
1. 将 Huffman 编码与行程编码结合,能否提高压缩效果?试验证之。
2. 行程编码是否可以改进?若能,应如何改进?
3. 试用行程编码方法对一幅真彩色图像压缩,用 PCX 格式存储。

实习十二　纹理图像的自相关函数分析法

【实习目的】
加深对图像纹理分析方法的理解与应用,掌握纹理图像的自相关函数分析方法,启发学生深入研究纹理分析方法,提高学生图像处理与分析能力和实际动手能力。

【实习内容】
1. 编程实现纹理图像的自相关函数分析方法。
2. 观察其周期性及其大小,分析纹理基元分布的疏密程度,识别纹理的粗细。

【预备知识】
1. 熟悉位图图像读取。
2. 熟悉 VC 编程。

【实习原理】
若一幅图像为 $f(i,j)$, $i,j=0,1,\cdots,N-1$,则该图像的自相关函数定义为

$$\rho(x,y) = \frac{\sum_{i=1}^{N-1}\sum_{j=0}^{N-1}f(i,j)f(i+x,j+y)}{\sum_{i=0}^{N-1}\sum_{j=0}^{N-1}f(i,j)^2} \tag{1}$$

$\rho(x,y)$ 可以看成一幅大小为 $N \times N$ 的图像。自相关函数 $\rho(x,y)$ 随 x,y 大小而变化,其变化与图像中纹理粗细的变化有着对应的关系,因而可描述图像纹理特征。定义 $d=(x^2,y^2)^{1/2}$,d 为位移矢量,$\rho(x,y)$ 可以记为 $\rho(d)$。在 $x=0$,$y=0$ 时,由自相关函数定义可以得出,$\rho(d)=1$ 为最大值。不同的纹理图像,$\rho(x,y)$ 随 d 变化的规律不同。当纹理较粗时,$\rho(d)$ 随 d 的增加下降速度较慢;当纹理较细时,$\rho(d)$ 随着 d 的增加下降速度较快。随着 d 的继续增加,$\rho(d)$ 则会呈现某种周期性的变化,其周期大小可以描述纹理基元分布的疏密程度。图 12-33 所示是计算自相关函数的流程图。

【实习步骤】
1. 打开上次实习所用的项目。
2. 在自己创建的类中添加计算自相关函数的操作。
3. 如图 12-34 所示添加菜单项"纹理特征提取/相关系数分析法"。
4. 为新加的菜单项建立消息处理函数。
5. 在相应函数体内添加实现根据自相关函数来统计纹理特征的源代码(通过类对象中的操作来实现),并根据统计 $\rho(d)$ 的变化规律来得出图像的纹理特性。

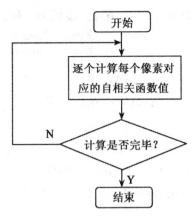
图 12-33 计算自相关函数的流程图

图 12-34 添加菜单项

6.编译检查语法错误。若编译通过,运行程序,观察、分析得到的纹理特性是否与实际情况相符合。

实习完毕后,提交一份实习报告。

【思考题】
能否利用自相关函数分析法进行纹理图像分割?若能,如何分割?

实习十三 灰度共生矩阵特征提取

【实习目的】
加深对图像纹理分析方法的理解与应用,掌握灰度共生矩阵特征提取与分析方法,启发学生深入研究纹理分析方法,提高学生图像处理与分析能力和实际动手能力。

【实习内容】
1. 编程计算灰度共生矩阵。
2. 由灰度共生矩阵提取常用特征,实现对图像纹理特性的分析。

【预备知识】
1. 熟悉 BMP 图像数据的读取。
2. 熟悉 VC 编程。
3. 熟悉灰度共生矩阵特征的提取方法。

【实习原理】
灰度共生矩阵是从图像灰度为 i 的像素 (x,y) 出发,统计与距离为 δ、灰度为 j 的像素 $(x+\Delta x, y+\Delta y)$ 同时出现的概率 $P(i,j,\delta,\theta)$,即

$$P(i,j,\delta,\theta) = \{[(x,y),(x+\Delta x, y+\Delta y)] \mid f(x,y)=i, f(x+\Delta x, y+\Delta y)=j; x=0,1,\cdots,N_x-1; y=0,1,\cdots,N_y-1\} \tag{1}$$

式中 $i,j=0,1,\cdots,L-1$;x,y 是图像中的像素坐标;L 为图像的灰度级数;N_x, N_y 分别是图像的

行列数。

一般在 0°、45°、90°和 135°四个方向上,计算相距为 δ、图像灰度值为 i 和 j 的两像素出现的概率为 $P(i,j,\delta,\theta)$。当 $\delta=1$ 时,则有

对于水平方向:

$$P(i,j,\delta,0°)=\left\{[(x,y),(x+\Delta x,y+\Delta y)]\left|\begin{array}{l}f(x,y)=i,f(x+\Delta x,y+\Delta y)=j;x=0,1,\cdots,N_x-1;\\ |\Delta x|=1,\Delta y=0 \quad y=0,1,\cdots,N_y-1\end{array}\right.\right\}$$
(2)

对于垂直方向:

$$P(i,j,\delta,90°)=\left\{[(x,y),(x+\Delta x,y+\Delta y)]\left|\begin{array}{l}f(x,y)=i,f(x+\Delta x,y+\Delta y)=j;x=0,1,\cdots,N_x-1;\\ \Delta x=0,|\Delta y|=1 \quad y=0,1,\cdots,N_y-1\end{array}\right.\right\}$$
(3)

对于 45°方向:

$$P(i,j,\delta,45°)=\left\{[(x,y),(x+\Delta x,y+\Delta y)]\left|\begin{array}{l}f(x,y)=i,f(x+\Delta x,y+\Delta y)=j;x=0,1,\cdots,N_x-1;\\ |\Delta x|=1,|\Delta y|=1 \quad y=0,1,\cdots,N_y-1\end{array}\right.\right\}$$
(4)

对于 135°方向

$$P(i,j,\delta,135°)=\left\{[(x,y),(x+\Delta x,y+\Delta y)]\left|\begin{array}{l}f(x,y)=i,f(x+\Delta x,y+\Delta y)=j;x=0,1,\cdots,N_x-1;\\ |\Delta x|=1,|\Delta y|=1 \quad y=0,1,\cdots,N_y-1\end{array}\right.\right\}$$
(5)

灰度共生矩阵反映了图像灰度关于方向、相邻间隔、变化幅度的综合信息,它可作为分析图像基元和排列结构的信息。作为纹理分析的特征量,往往不是直接应用计算的灰度共生矩阵,而是在灰度共生矩阵的基础上再提取纹理特征量,称为二次统计量。

一幅图像的灰度级数一般为 256,这样级数太多会导致计算的灰度共生矩阵大,计算量大。为了解决这一问题,在求灰度共生矩阵之前,先将图像灰度级数压缩为 16。这里要求用计算得到的灰度共生矩阵来提取常用的 5 个特征:

1. 二阶矩(能量)

$$f_1=\sum_{i=0}^{L-1}\sum_{j=0}^{L-1}p^2(i,j,\delta,\theta)$$
(6)

二阶矩反映了图像灰度分布均匀程度和纹理粗细度。因为它是灰度共生矩阵各元素的平方和,又称为能量。f_1 大时纹理粗,能量大;反之,f_1 小时纹理细,能量小。

2. 对比度(惯性矩)

$$f_2=\sum_{n=0}^{L-1}n^2\left\{\sum_{\substack{i=0\\n=|i-j|}}^{L-1}\sum_{j=0}^{L-1}p^2(i,j,\delta,\theta)\right\}$$
(7)

对比度可以理解为图像的清晰度。纹理的纹沟深,f_2 大,效果清晰;反之,f_2 小则纹沟浅,效果模糊。

3. 相关

$$f_3 = \frac{\sum_{i=0}^{L-1}\sum_{j=0}^{L-1} ijp(i,j,\delta,\theta) - u_1 u_2}{\sigma_1^2 \sigma_2^2} \tag{8}$$

式中，u_1，u_2，σ_1，σ_2 分别定义为：

$$u_1 = \sum_{i=0}^{L-1} i \sum_{j=0}^{L-1} p(i,j,\delta,\theta)$$

$$u_2 = \sum_{j=0}^{L-1} j \sum_{i=0}^{L-1} p(i,j,\delta,\theta)$$

$$\sigma_1^2 = \sum_{i=0}^{L-1} (i-u_1)^2 \sum_{j=0}^{L-1} p(i,j,\delta,\theta)$$

$$\sigma_2^2 = \sum_{j=0}^{L-1} (j-u_2)^2 \sum_{i=0}^{L-1} p(i,j,\delta,\theta)$$

相关用来衡量灰度共生矩阵在行或列方向上的相似度。例如，水平走向的纹理在 $\theta=0$ 方向上的 f_3 大于其他方向的 f_3。

4. 熵

$$f_4 = -\sum_{i=0}^{L-1}\sum_{j=0}^{L-1} p(i,j,\delta,\theta)\log_2 p(i,j,\delta,\theta) \tag{9}$$

它反映图像中纹理的复杂程度或非均匀度。若纹理复杂，熵具有较大值；反之，若图像中灰度均匀，共生矩阵中元素大小差异大，熵较小。

5. 逆差矩

$$f_5 = \sum_{i=0}^{L-1}\sum_{j=0}^{L-1} \frac{p(i,j,\delta,\theta)}{1+(i-j)^2} \tag{10}$$

若希望提取具有旋转不变性的特征，简单的方法是对 θ 取 0°、45°、90°、135°的同一特征求平均值和均方差。图 12-35 所示为灰度共生矩阵特征提取的流程。

【实习步骤】
1. 在自己创建的类中添加计算灰度共生矩阵以及以上 5 个特征提取的操作。
2. 添加菜单项"纹理特征提取/灰度共生矩阵分析法"。
3. 为新加的菜单项建立消息处理函数。
4. 在相应函数体内添加提取 5 个纹理特征的源代码。
5. 编译检查语法错误，若编译通过，运行程序，分析计算得到的纹理特性是否与实际情况相符合。

实习完毕后，提交一份实习报告。

【思考题】
1. 分析每个特征量所反映的纹理特性。
2. 灰度共生矩阵分析法与自相关函数分析法各有何特点？
3. 针对高分辨率遥感图像的不同类别区域，分别提取各类灰度共生矩阵特征并进行比较，分析利用纹理信息识别目标的可行性。

第三部分 综合设计

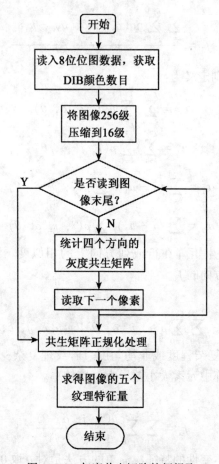

图 12-35　灰度共生矩阵特征提取

实习十四　基于灰度的模板匹配

【实习目的】
加深对图像目标识别方法的理解与应用，掌握模板匹配的基本算法，启发学生深入研究图像识别方法，提高学生图像处理与分析能力和实际动手能力。

【实习内容】
1. 编程实现基于灰度的模板匹配算法。
2. 分析该方法的适用性，编程实现快速算法。

【预备知识】
1. 熟悉图像局部处理的方法。
2. 熟悉 BMP 位图文件的读写和显示。

【实习原理】
所谓模板匹配，是根据模板与一幅图像的各部分的相似度判断其是否存在，并求得模板在图像中位置的操作。设模板 T 叠放在搜索图像 S 上平移,模板覆盖的搜索区叫做子图像 $S_{i,j}$,i,j 为子图像的左上角在 S 图像中的坐标,i,j 的取值范围为 $1<i$,$j<N-M+1$,如图 12-36 所示。

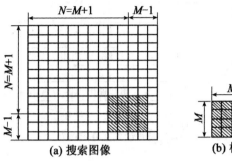

图 12-36 模板匹配

衡量子图像和模板之间的差别用平方误差之和或绝对差之和来表示,即

$$D(i, j) = \sum_{m=1}^{M} \sum_{n=1}^{M} [S_{i,j}(m, n) - T(m, n)]^2 \tag{1}$$

或者

$$D(i, j) = \sum_{m=1}^{M} \sum_{n=1}^{M} |S_{i,j}(m, n) - T(m, n)| \tag{2}$$

也可用相关函数作为相似性测度

$$R(i, j) = \frac{\sum_{m=1}^{M} \sum_{n=1}^{M} S_{i,j}(m, n) \times T(m, n)}{\sum_{m=1}^{M} \sum_{n=1}^{M} [S_{i,j}(m, n)]^2} \tag{3}$$

或者归一化为

$$R(i, j) = \frac{\sum_{m=1}^{M} \sum_{n=1}^{M} S_{i,j}(m, n) \times T(m, n)}{\sqrt{\left(\sum_{m=1}^{M} \sum_{n=1}^{M} [S_{i,j}(m, n)]^2\right)} \sqrt{\left(\sum_{m=1}^{M} \sum_{n=1}^{M} [T(m, n)]^2\right)}} \tag{4}$$

图 12-37 给出了基于灰度的模板匹配流程图。

【实习步骤】

1. 打开上次实习所用的项目。
2. 在自己创建的类中添加基于灰度的模板匹配操作。
3. 如图 12-38 所示添加菜单项"基于灰度的模板匹配"菜单项。
4. 为新加的菜单项建立消息处理函数。
5. 在相应函数体内添加实现基于灰度的模板匹配的源代码(通过类对象中的操作来实现)。
6. 编译检查语法错误,若编译通过,运行程序,观察基于灰度模板匹配得到的结果

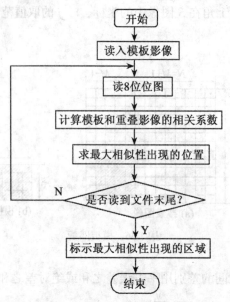

图 12-37 基于灰度的模板匹配流程图

图 12-38 添加菜单项

是否与模板一致。

实习完毕后,提交一份实习报告。

【思考题】

设计一种基于投影的模板匹配算法,同常规方法比较,验证方法的有效性。

实习十五 基于特征的模板匹配

【实习目的】

加深对图像目标识别方法的理解与应用,掌握基于特征的模板匹配的基本算法及其特性,启发学生深入研究图像识别方法,提高学生图像处理与分析能力和实际动手能力。

【实习内容】

1. 编程实现一种基于特征的模板匹配算法。
2. 同基于灰度的模板匹配法结果比较,分析算法的优缺点。

【预备知识】

1. 熟悉图像特征提取。

2. 熟悉图像局部处理方法。

3. 熟悉 BMP 位图文件的读写和显示。

【实习原理】

由于图像往往有较强自相关性，因此，进行模板匹配计算的相似度就在以对象物存在的地方为中心形成平缓的峰。这样，即使从图像中对象物的真实位置稍微离开一点，也表现出相当高的相似度。为了求得对象物的精确位置，总希望相似度分布尽可能尖锐一些。

为了达到这一目的，提出了基于轮廓特征的模板匹配方法。轮廓匹配与一般的匹配相比较，表现出更尖锐的相似度的分布。但其方法与基于灰度的模板匹配相似，只是这里通过对图像轮廓等形状特征进行匹配，从而提高匹配精度。流程图请参阅图 3-43 基于灰度的模板匹配。

【实习步骤】

1. 打开上次实习所用的项目。
2. 在自己创建的类中添加基于特征的模板匹配操作。
3. 添加菜单项"基于特征的模板匹配"菜单项。
4. 为新加的菜单项建立消息处理函数。
5. 在相应函数体内添加实现基于特征的模板匹配的源代码(通过类对象中的操作来实现)。
6. 编译检查语法错误，若编译通过，运行程序，观察基于特征模板匹配得到的结果是否与模板一致。

实习完毕后，提交一份实习报告。

【思考题】

设计一种基于形状特征的快速模板匹配算法用于识别影像规则形状目标或数字地图符号。

实习十六　形状特征提取

【实习目的】

熟悉二值图像增强处理和形状提取的基本方法，编程实现二值图像的区域特征参数提取功能，有利于依据形状特征进行目标识别。

【实习内容】

在新建的类中加入区域参数提取操作，使自己的工程能实现二值图像区域面积、周长以及圆形度的计算操作。

【预备知识】

1. 熟悉二值图像处理基本概念和方法。
2. 熟悉图像读写和显示。

【实习原理】

二值图像中区域的形状参数主要包括：

（1）面积：由区域内像素的总和确定。

（2）周长：常用计算方法有两种：一种是针对区域的边界像素而言，上、下、左、右像素间的距离为1，对角线像素间的距离为$\sqrt{2}$。周长就是边界像素间距离的总和。另一种是将边界的像素总和作为周长。

（3）圆形度：它是测量区域形状常用的量。其定义如下：

$$R = 4\pi \frac{\text{面积}^2}{\text{周长}} \tag{1}$$

当区域为圆形时，R最大（$R=1$）；如果是细长的区域，R则较小。

按图12-39的流程提取这些形状特征。

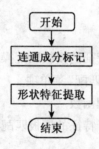

图12-39 提取形状特征的步骤

【实习步骤】

1. 打开上次实习所用的项目。
2. 在自己创建的类中添加"形状分析"的操作。
3. 添加菜单项"区域测量"，以及区域测量下的子菜单"面积"、"周长"和"圆形度"。
4. 为新加的菜单项建立消息处理函数。
5. 在相应函数体内添加源代码(通过类对象中的操作来实现)。
6. 编译检查语法错误，若编译通过，运行程序，观察计算结果是否与实际情况相符合。

实习完毕后，提交一份实习报告。

【思考题】

1. 编程实现二值图像中多目标区域的标号和几何特征提取。
2. 编程实现目标形状特征参数查询功能。

实习十七 色彩平衡

【实习目的】

熟悉彩色图像存在偏色的情况下，对图像进行色彩平衡处理的原理和方法，编程实现图像的偏色校正，即色彩平衡。提高学生的彩色图像处理技能。

【实习内容】

在新建的类中加入色彩平衡操作，使自己的工程能实现对偏色图像的正处理。

【预备知识】

1. 熟悉图像读写和显示。
2. 熟悉彩色图像处理基本概念和方法。

【实习原理】

当一幅彩色图像数字化后，因为颜色通道中不同的敏感度、增光因子、偏移量（黑级）等，导致数字化中的三个图像分量出现不同的变换，使结果图像的三原色"不平衡"，从而使影象中物体的颜色偏离了其原有的真实色彩。彩色平衡处理的目的就是将有色偏的图像进行颜色校正，获得正常颜色的图像。下面，介绍一种基本的彩色平衡处理方法，此方法是对白平衡方法的一种改进。

白平衡原理是，如果原始场景中的某些像素点应该是白色的（即 $R_k^* = G_k^* = B_k^* = 255$），但是由于所获得图像中的相应像素点存在色偏，这些点的 R，G，B 三个分量的值不再保持相同，通过调整这三个颜色分量的值，使之达到平衡，由此获得对整幅图像的彩色平衡映射关系，通过该映射关系对整幅图像进行处理，即可达到彩色平衡的目的。

方法的具体步骤如下：

(1) 对拍摄到的有色偏的图像，按照下式计算该图像的亮度分量。

$$Y = 0.299 \times R + 0.587 \times G + 0.114 \times B \tag{1}$$

获得图像的亮度信息 Y_{max} 和平均亮度 $\overline{Y}$。

(2) 考虑到对环境光照的适应性，寻找出图像中所有灰度值 $\leq 0.95 Y_{max}$ 像素点，将这些点假设为原始场景中的灰色点，即设这些点所构成的像素点集为灰色点集 $\{f(i,j) \in \Omega_{gray}\}$。

(3) 计算灰色点集 Ω_{gray} 中所有像素的 R，G，B 三个颜色分量的 $\overline{R}$，$\overline{G}$，$\overline{B}$。

(4) 求出 $\overline{R}$，$\overline{G}$，$\overline{B}$ 三个值中最大的值 $B_{max} = \max(\overline{R}, \overline{G}, \overline{B})$，然后计算颜色均衡系数 K_R，K_G，K_B，计算公式如下：

$$K_R = \frac{B_{max}}{\overline{R}}, \quad K_G = \frac{B_{max}}{\overline{G}}, \quad K_B = \frac{B_{max}}{\overline{B}} \tag{2}$$

(5) 对整幅图像的 R，G，B 三个颜色分量，进行彩色平衡调整如下：

$$R^* = K_R \cdot R, \quad G^* = K_G \cdot G, \quad B^* = K_B \cdot B \tag{3}$$

白平衡算法流程如图 12-40 所示。

【实习步骤】

1. 打开上次实习所用的项目。
2. 在自己创建的类中添加"色彩平衡"的操作。
3. 添加菜单项"色彩均衡处理"。
4. 为新加的菜单项建立消息处理函数。
5. 在相应函数体内添加源代码（通过类对象中的操作来实现）。

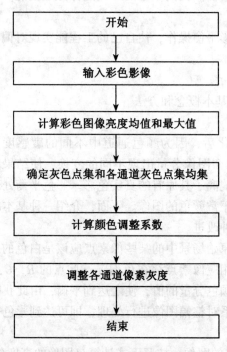

图 12-40　色彩平衡流程图

6. 编译检查语法错误，若编译通过，运行程序，观察计算计算结果是否与实际情况相符合。

实习完毕后，提交一份实习报告。

【思考题】

1. 分析给出的白平衡法，给出其适用范围。
2. 设计与实现一种新的色彩平衡法。

实习十八　点特征提取

【实习目的】

熟悉与掌握数字图像角点特征提取的基本方法。编程实现 SUSAN 角点提取算子，以便用于图像匹配、目标识别等。

【实习内容】

1. 编程实现基于 SUSAN 算子的角点提取算法。
2. 分析该算法的特点。

【预备知识】

1. 熟悉 BMP 位图文件的读写和显示方法。
2. 了解 SUSAN 算子角点提取的原理。
3. 熟悉数字图像的局部处理方法。

【实习原理】

用圆形模板在图像上移动，若模板内像素的灰度与模板中心像素（核）灰度的差值小于给定的门限，则认为该点与核具有相同（或相近）的灰度，由满足这样条件的像素组成的区域称为吸收核同值区，称为 USAN。

具体检测时，是用圆形模板扫描整个图像，比较模板内的每一像素与中心像素的灰度值，并给定阈值来判断该像素是否属于 USAN 区域。

USAN 区域的大小反映了图像局部特征的强度。当圆形模板完全处在背景或目标中时，USAN 区域面积最大；当模板移向目标边缘时，USAN 区域逐渐变小；当模板中心处于边缘时，USAN 区域很小；当模板中心处于角点时，USAN 区域最小。在得到每个像素的 USAN 区域后，再由下式产生 USAN 特征图像。

$$R(r_0) = \begin{cases} g - n(r_0), & n(r_0) < g \\ 0, & 其他 \end{cases} \tag{1}$$

式中，g 为阈值。g 取得越小，所检测到的角点越尖锐。采用这种方法，取不同的阈值，不但能检测特征点，还可以检测角点、边缘等特征。利用 SUSAN 算子提取角点的流程如图 12-41 所示。

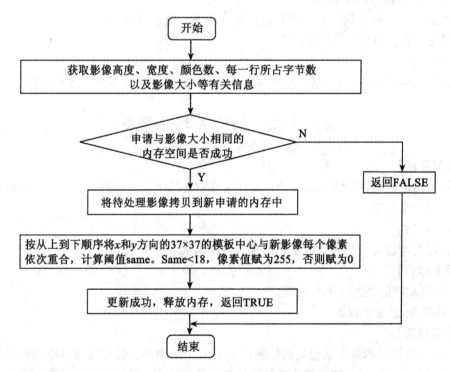

图 12-41 SUSAN 算子提取角点的流程

【实习步骤】

1. 打开上次实习所用的项目。
2. 在自己创建的类中添加"图像点特征提取"操作。

3. 添加"SUSAN 算子角点检测"子菜单，如图 12-42 所示。

图 12-42 添加菜单项

4. 为新加的菜单项建立消息处理函数。
5. 在相应函数体内添加源代码。
6. 编译检查语法错误，若编译通过，运行程序，观察处理结果。
实习完毕后，提交一份实习报告。

【思考题】
编程实现基于其他算子的点特征提取，与 SUSAN 算子的检测效果进行比较。

实习十九 图像 K 均值聚类

【实习目的】
掌握 K 均值聚类算法的基本原理，编程实现对遥感图像 K 均值分类算法，分析 K 均值聚类算法的分类性能。

【实习内容】
给自建类增加 K 均值聚类方法，编程实现图像 K 均值聚类功能。

【预备知识】
1. 熟悉 K 均值聚类的基本原理。
2. 熟悉 BMP 位图读取。

【实习原理】
（1）K 均值算法的基本思想是初始随机给定 K 个簇中心，按照最邻近原则将待分类的样本点分到各个簇，然后按平均法重新计算各个簇的质心，从而确定新的类别中心。一直迭代，直到簇心的移动距离小于某个给定的值，则停止迭代。K 均值算法要求各类样本到聚类中心的距离平方和最小，它是在误差平方和准则的基础上建立起来的。K 均值聚类算法流程图如图 12-43 所示。

K 均值聚类算法主要分为四个步骤：

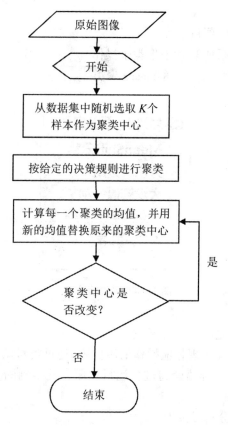

图 12-43 K 均值算法流程图

(1) 任选 K 个聚类中心 $Z_1(1)$, $Z_2(1)$, $Z_3(1)$, $\cdots$, $Z_k(1)$

(2) 逐个将模式样本集 $\{x\}$ 的每一样本按最小距离原则分配给 K 个聚类中心,形成 K 个类群,即在第 m 次迭代时,若

$$|| x - z_j(m) || < || x - z_i(m) ||, \ i = 1, 2, \cdots, K, \ i \neq j \tag{1}$$

则
$$x \in f_j(m) \tag{2}$$

式(2)中,$f_j(m)$ 表示第 m 次迭代时,以第 j 个聚类中心为代表的聚类域。

(3) 由步骤(2) 计算新的聚类中心,即

$$z_i(m+1) = \frac{1}{N_i} \sum_{x \in f_i(m)} x, \ i = 1, 2, \cdots, K \tag{3}$$

式中,N_i 为第 i 个聚类域的样本个数,其均值向量作为新的聚类中心,因为这样可以使得误差平方和准则函数

$$J = \sum_{x \in f_i(m)} || x - z_i(m+1) ||^2, \ i = 1, 2, \cdots, K \tag{4}$$

达到最小值。

(4) 若 $z_i(m+1) = z_i(m)$, $i = 1, 2, \cdots, K$, 算法收敛,计算完毕。否则回到步骤(2),进行下一次迭代。

【实习步骤】

1. 打开上次实习所用的项目。
2. 在自己创建的类中添加 K 均值聚类子操作。
3. 如图 12-44 所示添加菜单项"K-means 聚类"。

图 12-44　添加菜单项

4. 为新加的菜单项建立消息处理函数。
5. 在相应函数体内添加实现相应操作的源代码(通过类对象中的操作来实现)。
6. 编译检查语法错误。若编译通过，运行程序，观察判断 K-means 分类效果，对结果进行评价。

实习完毕后，提交一份实习报告。

【思考题】

1. 分析所实现的 K 均值聚类算法的效果及特点。
2. 试编程实现 ISODATA 算法对遥感影像进行分类，通过调节 ISODATA 人机交互环节中的 6 个参数来分析聚类参数对分类精度的影响。

实习二十　分水岭分割

【实习目的】

进一步掌握分水岭分割的原理，编程实现分水岭分割算法，提高学生图像处理与分析能力和实际动手能力。

【实习内容】

给自建类增加分水岭分割方法，编程实现分水岭分割功能。

【预备知识】

1. 熟悉分水岭分割的基本原理。
2. 熟悉 BMP 位图读取。

【实习原理】

分水岭分割(Watershed Segmentation)是一种结合了地形学和区域生长思想的图像分割算法。该算法把一幅灰度图像看作是地形图，灰度值高的像素区域代表高山，灰度值低的像素区域代表低洼。假设下雨时，水会顺着高山的侧面流向低洼处，形成"湖泊"，图像中的这些"湖泊"称为集水盆地。而当集水盆地中水位不断升高，水有可能会溢出流向其他周围的集水盆地，若在每个集水盆地交界处建立水坝，水就不会溢出。这些水坝的位置就是分水线，也就是我们要找的图像分割结果。将分水岭原理应用到数字图像中，就把一幅图像分为单个的小区域，水流的方向由图像的梯度值来判定，依据各个像素点的方向来直接确定系列的最小点值的区域，汇水盆地是的彼此不连通区域，基于数学形态学的分水岭法方法广泛的应用于图像分割领域。分水岭示意图如图12-45所示。

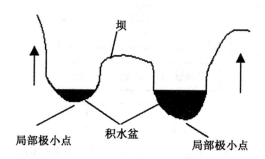

图 12-45　分水岭示意图

水岭分割法的主要应用是从背景中分割出感兴趣的前景对象。图像中灰度级上变化较小的区域的梯度值也较小。因此，分水岭分割方法与图像的梯度有很大的关系，而不是原图像，集水盆地的局部最小值对应梯度图像中目标对象的小的梯度值。分水线的构造是以二值图像为基础的形态学运算，属于二维整数空间 Z^2。构造分水线的最简单的方法是使用形态学膨胀运算。

基于标记的分水岭算法的主要思想是：使用强制最小值标定算法先确定梯度图像中的最小值区域，强制性地将提取的标记作为梯度图像的极小值，修改原梯度图像。然后在修改后的梯度图像基础上应用分水岭算法，完成图像分割。基于标记的分水岭分割算法采用内部标记和外部标记，一个标记就是一个联通成分，内部标记与某个感兴趣的目标相关，外部标记与背景相关。标记的选取包括预处理和定义一组选取标记的准则。标记选择准则可以是灰度值、连通性、尺寸、形状、纹理等特征。有了内部标记之后，就只以这些内部标记为极小值区域进行分割，分割结果的分水线作为外部标记，然后对每个分割出来的区域利用其他分割技术，例如阈值化，将背景与目标分离出来。

基于标记的分水岭分割算法步骤：
1. 将输入图像转化为灰度图像。
2. 求灰度图像的梯度图像。
3. 对灰度图像进行形态学开重建运算和形态学闭重建运算。
4. 对第3步得到的结果图像计算前景标记。
5. 对第3步得到的结果图像计算背景标记。
6. 根据前景标记和背景标记修改梯度图像，进行分水岭分割。
7. 得到分割结果。

图12-46所示为基于标记的分水岭分割的流程图。

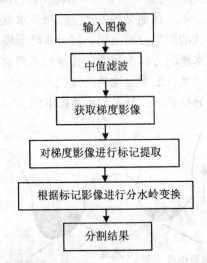

图12-46 基于标记的分水岭分割流程图

【实习步骤】
1. 打开上次实习所用的项目。
2. 在自己创建的类中添加各种图像分割子操作。
3. 如图12-47所示添加菜单项"分水岭分割"。

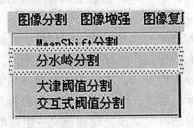

图12-47 添加菜单项

4. 为新加的菜单项建立消息处理函数。
5. 在相应函数体内添加实现相应操作的源代码(通过类对象中的操作来实现)。
6. 编译检查语法错误。若编译通过，则运行程序，观察判断分水岭分割法分割效果，对结果进行评价。

实习完毕后，提交一份实习报告。
【思考题】
分析所实现的分水岭分割算法效果及特点。

第四部分 实习项目的 C++源程序代码

《数字图像处理实习教程(第三版)》前面章节中已经给出了图像的读取、存储和显示源代码,并且在前面实习教程中已经给出了 RAW→BMP 转换的源代码。因此,为了避免重复,这里只给出了其余实习项目的有关 C++源程序代码供参考。该源程序是在第二部分第六章通用平台下设计实现的。

1. 线性拉伸

```
/***************************************************************
灰度线性变换/拉伸
参数:
                    int minout        线性变化下限
                    int maxout        线性变化上限
返回:               BOOL
***************************************************************/
BOOL LinearTransform(int minout, int maxout)
{
    //判断图像数据类型
    switch(datatype)
    {
    case GDT_Byte:
        LinearTransform(byte(0), minout, maxout);
        break;
    case GDT_UInt16:
        LinearTransform(WORD(0), minout, maxout);
        break;
    case GDT_Int16:
        LinearTransform(short(0), minout, maxout);
        break;
    case GDT_UInt32:
        LinearTransform(DWORD(0), minout, maxout);
        break;
    case GDT_Int32:
        LinearTransform(int(0), minout, maxout);
        break;
    case GDT_Float32:
        LinearTransform(float(0), minout, maxout);
        break;
    case GDT_Float64:
        LinearTransform(double(0), minout, maxout);
        break;
```

```cpp
        default:
            break;
    }
    return TRUE;
}
template <class T>
BOOL LinearTransform(T, int minout, int maxout)
{
    //判断图像是否为空
    if(! IsValid())
    {
        return FALSE;
    }
    //获取图像数据指针
    T * lpData = (T * )pData;
    //获取显示数据指针
    BYTE * pByte = m_pDIBs;

    //获取图像的宽度与高度
    DWORD nWidth = GetWidth();
    DWORD nHeight = GetHeight();

    int i, j, k;
    //逐波段进行处理
    for (k=0; k<nBands; k++)
    {
        //统计原图像最大和最小灰度级
        float m_MinGray=(float)999999999, m_MaxGray=(float)-99999999;
        for (i=0; i<nHeight; i++)
        {
            for (j=0; j<nWidth; j++)
            {
                if( * (lpData+nWidth * i+j+k * nWidth * nHeight)<m_MinGray)
                    m_MinGray = * (lpData+nWidth * i+j+k * nWidth * nHeight);
                if( * (lpData+nWidth * i+j+k * nWidth * nHeight)>m_MaxGray)
                    m_MaxGray = * (lpData+nWidth * i+j+k * nWidth * nHeight);
            }
        }
```

```
        //对源数据进行拉伸
        for(i=0; i<nHeight; i++)
        {
            for(j=0; j<nWidth; j++)
            {
                *(lpData+nWidth*i+j+k*nWidth*nHeight)=(T)(minout+(maxout-minout)/(m_MaxGray-m_MinGray)*(*(lpData+nWidth*i+j+k*nWidth*nHeight)-m_MinGray));
            }
        }
    }
    //修改显示数据，更新视图
    for(k = 0; k < nBandsShow; k++)
    {
        for(i=0; i<nHeight; i++)
        {
            for(j=0; j<nWidth; j++)
            {
                if(nBandsShow == 1)
                    *(pByte+lLineBYTES*(nHeight-1-i)+j*nBandsShow+k)=(BYTE)(*(lpData+nWidth*i+j+k*nWidth*nHeight));
                else
                    *(pByte+lLineBYTES*(nHeight-1-i)+j*3+2-k)=(BYTE)(*(lpData+nWidth*i+j+k*nWidth*nHeight));
            }
        }
    }
    return TRUE;
}
```

2. 直方图均衡化

```
/****************************************************************
直方图均衡化
参数：      无
返回：      BOOL
****************************************************************/
BOOL HistoEquivalize()
{
    //判断图像数据类型
    switch(datatype)
```

```cpp
    }
    case GDT_Byte:
        HistoEquivalize(byte(0));
        break;
    case GDT_UInt16:
        HistoEquivalize(WORD(0));
        break;
    case GDT_Int16:
        HistoEquivalize(short(0));
        break;
    case GDT_UInt32:
        HistoEquivalize(DWORD(0));
        break;
    case GDT_Int32:
        HistoEquivalize(int(0));
        break;
    case GDT_Float32:
        HistoEquivalize(float(0));
        break;
    case GDT_Float64:
        HistoEquivalize(double(0));
        break;
    default:
        break;
    }
    return TRUE;
}

template <class T>
BOOL HistoEquivalize(T)
{
    //判断图像是否为空
    if(! IsValid())
    {
        return FALSE;
    }
    //获取图像数据指针
    T * lpData = (T *)pData;
```

```cpp
//获取显示数据指针
BYTE *pByte = m_pDIBs;
//获取图像的宽度与高度
DWORD nWidth = GetWidth();
DWORD nHeight = GetHeight();
int i, j, k;
//逐波段进行处理
for (k=0; k<nBandsShow; k++)
{
    int r[256] = {0};
    //计算图像的最大灰度级
    int nMaxPixel = 0;
    for (i = 0; i < nHeight; i++)
    {
        for (j = 0; j < nWidth; j++)
        {
            if ( *(lpData+nWidth*i+j+k*nWidth*nHeight) > nMaxPixel)
            {
                nMaxPixel = *(lpData+nWidth*i+j+k*nWidth*nHeight);
            }
        }
    }
    //最大灰度级比最大灰度值至少要大1(如256级的最大值为255)
    nMaxPixel = (nMaxPixel+1<256)? 256：nMaxPixel+1;
    for(i=0; i<nHeight; i++)
    {
        for(j=0; j<nWidth; j++)
        {
            r[(int)(*(lpData+nWidth*i+j+k*nWidth*nHeight)*256/nMaxPixel)] += 1.0;
        }
    }
    //计算均衡化后各亮度的新值
    double percentile[256];
    for(i=0; i<256; i++)
    {
        percentile[i]=0.0;
        for(j=0; j<=i; j++)
```

```
                    {
                        percentile[i] += r[j];
                    }
                    percentile[i] *= (255. 0/(double)(nWidth * nHeight));
                }
                for (i=0; i<nHeight; i++)
                {
                    for (j=0; j<nWidth; j++)
                    {
  *(lpData+nWidth * i+j+k * nWidth * nHeight) = (T)percentile[(int)(*(lpData+i *
nWidth+j+k * nWidth * nHeight) * 256/nMaxPixel)];
                    }
                }
            }
        //修改显示数据,更新视图
        for (k = 0; k < nBandsShow; k++)
        {
            for (i=0; i<nHeight; i++)
            {
                for (j=0; j<nWidth; j++)
                {
                    if (nBandsShow == 1)
  *(pByte+lLineBYTES * (nHeight-1-i)+j * nBandsShow+k) = (BYTE)(*(lpData+
nWidth * i+j+k * nWidth * nHeight));
                    else
  *(pByte+lLineBYTES * (nHeight-1-i)+j * 3+2-k) = (BYTE)(*(lpData+nWidth * i+j+k
 * nWidth * nHeight));
                }
            }
        }
        return TRUE;
}
```

3.3 * 3 低通滤波

```
/ *************************************************************
3 * 3 低通滤波
参数:        void
返回:        BOOL
*************************************************************/
```

```cpp
BOOL LowPassImage( )
{
    //判断图像数据类型
    switch ( datatype )
    {
    case GDT_Byte:
        LowPassImage( byte(0) );
        break;
    case GDT_UInt16:
        LowPassImage ( WORD(0) );
        break;
    case GDT_Int16:
        LowPassImage ( short(0) );
        break;
    case GDT_UInt32:
        LowPassImage ( DWORD(0) );
        break;
    case GDT_Int32:
        LowPassImage ( int(0) );
        break;
    case GDT_Float32:
        LowPassImage ( float(0) );
        break;
    case GDT_Float64:
        LowPassImage ( double(0) );
        break;
    default:
        break;
    }
    return TRUE;
}
template <class T>
BOOL LowPassImage(T)
{
    //判断图像是否为空
    if ( ! IsValid( ) )
    {
        return FALSE;
```

```cpp
    }
    //获取图像数据指针
    T *lpData = (T*)pData;
    //获取图像的宽度与高度
    DWORD nWidth = GetWidth();
    DWORD nHeight = GetHeight();

    T *lpDataCopy = new T[nHeight*nWidth*nBands];
    //获取显示数据指针
    BYTE *pByte = m_pDIBs;
    //拷贝源数据
    memcpy(lpDataCopy, lpData, nWidth*nHeight*nBands*sizeof(T));

    //图像变换开始
    int i, j, k;
    //定义9个指针变量,对应3*3模板的9个像素的存储地址
    T  *p1, *p2, *p3, *p4, *p5, *p6, *p7, *p8, *p9;
    //逐波段进行处理
    for (k=0; k<nBands; k++)
    {
        for(i=1; i<nHeight-1; i++)
            for(j=1; j<nWidth-1; j++)
            {
                //得到模板每个像素的地址,因滤波运算是在原位图基础上进行的,不能
                //使用变换后的像素灰度作滤波运算,这是开始时同学们容易犯的错误
                p1 = lpDataCopy+nWidth*(i-1)+(j-1)+k*nWidth*nHeight;
                p2 = lpDataCopy+nWidth*(i)+(j-1)+k*nWidth*nHeight;
                p3 = lpDataCopy+nWidth*(i+1)+(j-1)+k*nWidth*nHeight;
                p4 = lpDataCopy+nWidth*(i-1)+(j)+k*nWidth*nHeight;
                //使p5指针指向原始影像的模板中心对应像素
                p5 = lpData+nWidth*(i)+(j)+k*nWidth*nHeight;
                p6 = lpDataCopy+nWidth*(i+1)+(j)+k*nWidth*nHeight;
                p7 = lpDataCopy+nWidth*(i-1)+(j+1)+k*nWidth*nHeight;
                p8 = lpDataCopy+nWidth*(i)+(j+1)+k*nWidth*nHeight;
                p9 = lpDataCopy+nWidth*(i+1)+(j+1)+k*nWidth*nHeight;
                //得到模板中心对应原位图的灰度值
                *p5 = *(lpDataCopy+nWidth*(i)+(j)+k*nWidth*nHeight);
```

```
            //计算模板中心像素的灰度值
            if (datatype == GDT_Byte)
            {
                int t=(int)((*p1+*p2+*p3+*p4+*p5+*p6+*p7+*p8+*p9)/9);
                if (t > 255) *p5 = 255;
                else if (t < 0) *p5 = 0;
                else *p5 = (BYTE)t;
            }
            else
                *p5=(T)((*p1+*p2+*p3+*p4+*p5+*p6+*p7+*p8+*p9)/9);
        }
}
//修改显示数据，更新视图
for (k=0; k<nBandsShow; k++)
{
    //统计原图像最大和最小灰度级
    float m_MinGray=(float)999999999, m_MaxGray=(float)-99999999;
    if (datatype == GDT_Byte)
    {
        m_MaxGray = 255;
        m_MinGray = 0;
    }
    else
    {
        for (i=0; i<nHeight; i++)
        {
            for (j=0; j<nWidth; j++)
            {
                if(*(lpData+nWidth*i+j+k*nWidth*nHeight)<m_MinGray)
                    m_MinGray = *(lpData+nWidth*i+j+k*nWidth*nHeight);
                if(*(lpData+nWidth*i+j+k*nWidth*nHeight)>m_MaxGray)
                    m_MaxGray = *(lpData+nWidth*i+j+k*nWidth*nHeight);
            }
        }
    }
    for (i=0; i<nHeight; i++)
    {
        for (j=0; j<nWidth; j++)
```

```
                    if ( nBandsShow = = 1 )
         * ( pByte+lLineBYTES * ( nHeight-1-i ) +j ) = ( BYTE ) ( ( * ( lpData+nWidth * i+j+k *
nWidth * nHeight ) -m_ MinGray ) * 255/( m_ MaxGray-m_ MinGray ) ) ;
                    else
         * ( pByte+lLineBYTES * ( nHeight-1-i ) +j * 3+2-k ) = ( BYTE ) ( ( * ( lpData+nWidth * i+j
+k * nWidth * nHeight ) -m_ MinGray ) * 255/( m_ MaxGray-m_ MinGray ) ) ;
              }
         }
    }
    delete [ ]lpDataCopy;
    return TRUE;
}
```

4. 3 * 3 高通滤波

```
/ *****************************************************************
3 * 3 高通滤波
参数：           void
返回：           BOOL
******************************************************************/
BOOL HighPassImage( )
{
    //判断图像数据类型
    switch ( datatype )
    {
    case GDT_ Byte:
        HighPassImage ( byte(0) ) ;
        break;
    case GDT_ UInt16:
        HighPassImage ( WORD(0) ) ;
        break;
    case GDT_ Int16:
        HighPassImage ( short(0) ) ;
        break;
    case GDT_ UInt32:
        HighPassImage ( DWORD(0) ) ;
        break;
    case GDT_ Int32:
        HighPassImage ( int(0) ) ;
```

```
            break;
        case GDT_Float32:
            HighPassImage(float(0));
            break;
        case GDT_Float64:
            HighPassImage(double(0));
            break;
        default:
            break;
        }
        return TRUE;
}
template <class T>
BOOL HighPassImage(T)
{
    //判断图像是否为空
    if(! IsValid())
    {
        return FALSE;
    }
    //获取图像数据指针
    T *lpData = (T *)pData;
    //获取图像的宽度与高度
    DWORD nWidth = GetWidth();
    DWORD nHeight = GetHeight();
    //分配内存
    T *lpDataCopy = new T[nHeight * nWidth * nBands];
    //获取显示数据指针
    BYTE *pByte = m_pDIBs;
    //拷贝源数据
    memcpy(lpDataCopy, lpData, nWidth * nHeight * nBands * sizeof(T));
    //图像变换开始
    int i, j, k;
    //定义9个指针变量,对应3*3模板的9个像素的存储地址
    T *p1, *p2, *p3, *p4, *p5, *p6, *p7, *p8, *p9;
    //逐波段进行处理
    for(k=0; k<nBands; k++)
    {
```

```
for(i=1; i<nHeight-1; i++)
    for(j=1; j<nWidth-1; j++)
    {
        //得到模板每个像素的地址,因滤波运算是在原位图基础上进行的,不能
        //使用变换后的像素灰度作滤波运算,这是开始时同学们容易犯的错误
        p1 = lpDataCopy+nWidth*(i-1)+(j-1)+k*nWidth*nHeight;
        p2 = lpDataCopy+nWidth*(i)+(j-1)+k*nWidth*nHeight;
        p3 = lpDataCopy+nWidth*(i+1)+(j-1)+k*nWidth*nHeight;
        p4 = lpDataCopy+nWidth*(i-1)+(j)+k*nWidth*nHeight;
        //使p5指针指向原始影像的模板中心对应像素
        p5 = lpData+nWidth*(i)+(j)+k*nWidth*nHeight;
        p6 = lpDataCopy+nWidth*(i+1)+(j)+k*nWidth*nHeight;
        p7 = lpDataCopy+nWidth*(i-1)+(j+1)+k*nWidth*nHeight;
        p8 = lpDataCopy+nWidth*(i)+(j+1)+k*nWidth*nHeight;
        p9 = lpDataCopy+nWidth*(i+1)+(j+1)+k*nWidth*nHeight;
        //得到模板中心对应原位图的灰度值
        *p5 = *(lpDataCopy+nWidth*(i)+(j)+k*nWidth*nHeight);

        //计算模板中心像素的灰度值
        if (datatype == GDT_Byte)
        {
            int t=(int)(*p5*9-*p1-*p2-*p3-*p4-*p6-*p7-*p8-*p9);
            if (t > 255) *p5 = 255;
            else if (t < 0) *p5 = 0;
            else *p5 = (BYTE)t;
        }
        else
        {
            int t=(int)(*p5*9-*p1-*p2-*p3-*p4-*p6-*p7-*p8-*p9);
            if (t < 0)
            {
                *p5 = 0;
            }
            else
                *p5 = t;
        }
    }
}
```

```cpp
//修改显示数据，更新视图
for (k=0; k<nBandsShow; k++)
{
    //统计原图像最大和最小灰度级
    float m_MinGray=(float)999999999, m_MaxGray=(float)-99999999;
    if (datatype == GDT_Byte)
    {
        m_MaxGray = 255;
        m_MinGray = 0;
    }
    else
    {
        for (i=0; i<nHeight; i++)
        {
            for (j=0; j<nWidth; j++)
            {
                if( *(lpData+nWidth*i+j+k*nWidth*nHeight)<m_MinGray)
                    m_MinGray = *(lpData+nWidth*i+j+k*nWidth*nHeight);
                if( *(lpData+nWidth*i+j+k*nWidth*nHeight)>m_MaxGray)
                    m_MaxGray = *(lpData+nWidth*i+j+k*nWidth*nHeight);
            }
        }
    }
    for (i=0; i<nHeight; i++)
    {
        for (j=0; j<nWidth; j++)
        {
            if (nBandsShow == 1)
    *(pByte+lLineBYTES*(nHeight-1-i)+j)=(BYTE)((*(lpData+nWidth*i+j+k*nWidth*nHeight)-m_MinGray)*255/(m_MaxGray-m_MinGray));
            else
    *(pByte+lLineBYTES*(nHeight-1-i)+j*3+2-k)=(BYTE)((*(lpData+nWidth*i+j+k*nWidth*nHeight)-m_MinGray)*255/(m_MaxGray-m_MinGray));
        }
    }
}
delete []lpDataCopy;
return TRUE;
```

}

5. 中值滤波

/**

中值滤波

parameter： void

return： BOOL

**/

```cpp
BOOL MedianFilterImage( )
{
    //判断图像数据类型
    switch（datatype）
    {
    case GDT_Byte：
        MedianFilterImage(byte(0))；
        break；
    case GDT_UInt16：
        MedianFilterImage (WORD(0))；
        break；
    case GDT_Int16：
        MedianFilterImage (short(0))；
        break；
    case GDT_UInt32：
        MedianFilterImage (DWORD(0))；
        break；
    case GDT_Int32：
        MedianFilterImage (int(0))；
        break；
    case GDT_Float32：
        MedianFilterImage (float(0))；
        break；
    case GDT_Float64：
        MedianFilterImage (double(0))；
        break；
    default：
        break；
    }
    return TRUE；
}
```

```cpp
template <class T>
BOOL MedianFilterImage (T)
{
    //判断图像是否为空
    if ( ! IsValid( ) )
    {
        return FALSE;
    }
    //获取图像数据指针
    T *lpData = (T *)pData;
    //获取图像的宽度与高度
    DWORD nWidth = GetWidth( );
    DWORD nHeight = GetHeight( );
    //分配内存
    T *lpDataCopy = new T[nHeight * nWidth * nBands];
    //获取显示数据指针
    BYTE *pByte = m_pDIBs;
    //拷贝源数据
    memcpy(lpDataCopy, lpData, nWidth * nHeight * nBands * sizeof(T));
    //图像变换开始
    int i, j, k;
    //定义25个指针变量，对应5*5模板的25个像素的存储地址
    T *p1, *p2, *p3, *p4, *p5, *p6, *p7, *p8, *p9, *p10,
      *p11, *p12, *p13, *p14, *p15, *p16, *p17, *p18, *p19, *p20,
      *p21, *p22, *p23, *p24, *p25;
    //逐波段进行处理
    for (k=0; k<nBands; k++)
    {
        for(i=2; i<nHeight-2; i++)
        {
            for(j=2; j<nWidth-2; j++)
            {
                p1= lpDataCopy +nWidth * (i-2)+(j-2)+k * nWidth * nHeight;
                p2= lpDataCopy +nWidth * (i-1)+(j-2)+k * nWidth * nHeight;
                p3= lpDataCopy +nWidth * (i)+(j-2)+k * nWidth * nHeight;
                p4= lpDataCopy +nWidth * (i+1)+(j-2)+k * nWidth * nHeight;
                p5= lpDataCopy +nWidth * (i+2)+(j-2)+k * nWidth * nHeight;
```

p6 = lpDataCopy +nWidth * (i-2)+(j-1)+k * nWidth * nHeight;
p7 = lpDataCopy +nWidth * (i-1)+(j-1)+k * nWidth * nHeight;
p8 = lpDataCopy +nWidth * (i)+(j-1)+k * nWidth * nHeight;
p9 = lpDataCopy +nWidth * (i+1)+(j-1)+k * nWidth * nHeight;
p10 = lpDataCopy +nWidth * (i+2)+(j-1)+k * nWidth * nHeight;
p11 = lpDataCopy +nWidth * (i-2)+(j)+k * nWidth * nHeight;
p12 = lpDataCopy +nWidth * (i-1)+(j)+k * nWidth * nHeight;
//使p13指针指向原始影像的模板中心对应像素
p13 = lpData+nWidth * (i)+(j)+k * nWidth * nHeight;

p14 = lpDataCopy +nWidth * (i+1)+(j)+k * nWidth * nHeight;
p15 = lpDataCopy +nWidth * (i+2)+(j)+k * nWidth * nHeight;
p16 = lpDataCopy +nWidth * (i-2)+(j+1)+k * nWidth * nHeight;
p17 = lpDataCopy +nWidth * (i-1)+(j+1)+k * nWidth * nHeight;
p18 = lpDataCopy +nWidth * (i)+(j+1)+k * nWidth * nHeight;
p19 = lpDataCopy +nWidth * (i+1)+(j+1)+k * nWidth * nHeight;
p20 = lpDataCopy +nWidth * (i+2)+(j+1)+k * nWidth * nHeight;
p21 = lpDataCopy +nWidth * (i-2)+(j+2)+k * nWidth * nHeight;
p22 = lpDataCopy +nWidth * (i-1)+(j+2)+k * nWidth * nHeight;
p23 = lpDataCopy +nWidth * (i)+(j+2)+k * nWidth * nHeight;
p24 = lpDataCopy +nWidth * (i+1)+(j+2)+k * nWidth * nHeight;
p25 = lpDataCopy +nWidth * (i+2)+(j+2)+k * nWidth * nHeight;
//得到模板中心对应原位图的灰度值
*p13= *(lpDataCopy+nWidth * (i)+(j)+k * nWidth * nHeight);
T t, temp[25];

temp[0] = *p1;   temp[1] = *p2;   temp[2] = *p3;
temp[3] = *p4;   temp[4] = *p5;   temp[5] = *p6;
temp[6] = *p7;   temp[7] = *p8; temp[8]   = *p9;
temp[9] = *p10;   temp[10] = *p11;   temp[11] = *p12;
temp[12] = *p13; temp[13] = *p14; temp[14] = *p15;
temp[15] = *p16;   temp[16] = *p17; temp[17] = *p18;
temp[18] = *p19;   temp[19] = *p20;   temp[24] = *p25;

//起泡法排序
for (int m=1; m<25; m++)
{
    for(int n=m+1; n<25; n++)

```cpp
                    {
                        if(temp[n]>temp[m])
                        {
                            t=temp[m];
                            temp[m]=temp[n];
                            temp[n]=t;
                        }
                    }
                }
                *p13 = temp[12];
            }
        }
    }
}
//修改显示数据,更新视图
for (k=0; k<nBandsShow; k++)
{
    //统计原图像最大和最小灰度级
    float m_MinGray=(float)999999999, m_MaxGray=(float)-99999999;
    if (datatype == GDT_Byte)
    {
        m_MaxGray = 255;
        m_MinGray = 0;
    }
    else
    {
        for (i=0; i<nHeight; i++)
        {
            for (j=0; j<nWidth; j++)
            {
                if(*(lpData+nWidth*i+j+k*nWidth*nHeight)<m_MinGray)
                    m_MinGray=*(lpData+nWidth*i+j+k*nWidth*nHeight);
                if(*(lpData+nWidth*i+j+k*nWidth*nHeight)>m_MaxGray)
                    m_MaxGray=*(lpData+nWidth*i+j+k*nWidth*nHeight);
            }
        }
    }
    for (i=0; i<nHeight; i++)
    {
```

```
            for (j=0; j<nWidth; j++)
            {
                if (nBandsShow == 1)
    *(pByte+lLineBYTES*(nHeight-1-i)+j)=(BYTE)((*(lpData+nWidth*i+j+k
*nWidth*nHeight)-m_MinGray)*255/(m_MaxGray-m_MinGray));
                else
    *(pByte+lLineBYTES*(nHeight-1-i)+j*3+2-k)=(BYTE)((*(lpData+nWidth*i+j
+k*nWidth*nHeight)-m_MinGray)*255/(m_MaxGray-m_MinGray));
            }
        }
    }
    delete []lpDataCopy;
    return TRUE;
}
```

6. 伪彩色编码

```
/****************************************************************
伪彩色编码
参数：      void
返回：      void
****************************************************************/

void OnColor()
{
    //伪彩色编码
    //获取文档
    CImageProcessExDoc * pDoc = GetDocument();
    if(pDoc->m_Img.bitinfor->bmiHeader.biBitCount != 8)
    {
        MessageBox("伪彩色变换只处理8位图像!","系统提示", MB_OK);
        return;
    }

    //保存用户选择的伪彩色编码表索引
    int     nColor;

    //参数对话框
    CDlgColor dlgPara;
```

```cpp
//初始化变量值
if (pDoc->m_nColorIndex >= 0){
    // 初始选中当前的伪彩色
    dlgPara.m_nColor = pDoc->m_nColorIndex;
}
else{
    // 初始选中灰度伪彩色编码表
    dlgPara.m_nColor = 0;
}

//指向名称数组的指针
dlgPara.m_lpColorName = (LPSTR)ColorScaleName;

//伪彩色编码数目
dlgPara.m_nColorCount = COLOR_SCALE_COUNT;

//名称字符串长度
dlgPara.m_nNameLen = sizeof(ColorScaleName) / COLOR_SCALE_COUNT;

//显示对话框,提示用户设定平移量
if (dlgPara.DoModal() != IDOK){
//返回
    return;
}

//获取用户的设定
nColor = dlgPara.m_nColor;

//删除对话框
delete dlgPara;

//更改光标形状
BeginWaitCursor();

//判断伪彩色编码是否改动
if (pDoc->m_nColorIndex != nColor){
    // 调用ReplaceColorPal()函数变换调色板
    pDoc->m_Img.ReplaceDIBColorTable((LPBYTE)ColorsTable[nColor]);
```

```cpp
            // 更新调色板
            pDoc->m_Img.CreatePalette();

            // 更新类成员变量
            pDoc->m_nColorIndex = nColor;

            // 设置脏标记
            pDoc->SetModifiedFlag(TRUE);

            // 修改显示数据，修改显示数据，更新视图
            pDoc->UpdateAllViews(NULL);
        }

        //恢复光标
        EndWaitCursor();
    }
```

7. 图像平移

```
/*****************************************************************
图像平移
参数：
            DWORD dXOffset,         水平位移
            DWORD dYOffset          垂直位移
返回：      BOOL
*****************************************************************/
```

```cpp
BOOL TranslationImage(DWORD dXOffset, DWORD dYOffset)
{
    //判断图像数据类型
    switch(datatype)
    {
    case GDT_Byte:
        TranslationImage(byte(0), dXOffset, dYOffset);
        break;
    case GDT_UInt16:
        TranslationImage(WORD(0), dXOffset, dYOffset);
        break;
    case GDT_Int16:
        TranslationImage(short(0), dXOffset, dYOffset);
```

```cpp
            break;
        case GDT_UInt32:
            TranslationImage(DWORD(0), dXOffset, dYOffset);
            break;
        case GDT_Int32:
            TranslationImage(int(0), dXOffset, dYOffset);
            break;
        case GDT_Float32:
            TranslationImage(float(0), dXOffset, dYOffset);
            break;
        case GDT_Float64:
            TranslationImage(double(0), dXOffset, dYOffset);
            break;
        default:
            break;
    }
    return TRUE;
}
template <class T>
BOOL TranslationImage(T, DWORD dXOffset, DWORD dYOffset)
{
    //判断图像是否为空
    if(! IsValid())
    {
        return FALSE;
    }
    //获取图像数据指针
    T * lpData = (T *)pData;
    //获取图像的宽度与高度
    DWORD nWidth = GetWidth();
    DWORD nHeight = GetHeight();
    //分配内存
    T * lpDataCopy = new T[nHeight * nWidth * nBands];
    //获取显示数据指针
    BYTE * pByte = m_pDIBs;
    //拷贝源数据
    memcpy(lpDataCopy, lpData, nWidth * nHeight * nBands * sizeof(T));
    int i, j, k;
```

```cpp
//像素在源影像中的坐标
long i0 = 0;
long j0 = 0;
//逐波段进行处理
for (k=0; k<nBands; k++)
{
    for(i = 0; i < nHeight; i++)
    {
        for(j = 0; j < nWidth; j++)
        {
            T *lpTemp= lpData+nWidth*i+j+k*nWidth*nHeight;

            // 计算该像素在源影像中的坐标
            i0 = i - dXOffset;
            j0 = j - dYOffset;

            // 判断是否在源图范围内
            if((j0 >= 0) && (j0 < (long)nWidth) && (i0 >= 0) && (i0 < (long)nHeight))
            {
                *lpTemp = *(lpDataCopy +nWidth*(i0)+(j0)+k*nWidth*nHeight);
            }
            else
            {
                // 对于源图中没有的像素，直接赋值为0
                *lpTemp = 0;
            }
        }
    }
}
//修改显示数据，更新视图
for (k=0; k<nBandsShow; k++)
{
    //统计原图像最大和最小灰度级
    float m_MinGray=(float)999999999, m_MaxGray=(float)-99999999;
    if (datatype == GDT_Byte)
    {
```

```
            m_MaxGray = 255;
            m_MinGray = 0;
        }
        else
        {
            for (i=0; i<nHeight; i++)
            {
                for (j=0; j<nWidth; j++)
                {
                    if( *(lpData+nWidth*i+j+k*nWidth*nHeight)<m_MinGray)
                        m_MinGray = *(lpData+nWidth*i+j+k*nWidth*nHeight);
                    if( *(lpData+nWidth*i+j+k*nWidth*nHeight)>m_MaxGray)
                        m_MaxGray = *(lpData+nWidth*i+j+k*nWidth*nHeight);
                }
            }
        }
        for (i=0; i<nHeight; i++)
        {
            for (j=0; j<nWidth; j++)
            {
                if (nBandsShow == 1)
    *(pByte+lLineBYTES*(nHeight-1-i)+j)=(BYTE)((*(lpData+nWidth*i+j+k
*nWidth*nHeight)-m_MinGray)*255/(m_MaxGray-m_MinGray));
                else
    *(pByte+lLineBYTES*(nHeight-1-i)+j*3+2-k)=(BYTE)((*(lpData+nWidth*i+j
+k*nWidth*nHeight)-m_MinGray)*255/(m_MaxGray-m_MinGray));
            }
        }
    }
    delete []lpDataCopy;
    return TRUE;
}
```

8. 图像缩放

/**

图像缩放
参数:
    float fXZoomRatio   X 轴方向缩放比率
    float fYZoomRatio   Y 轴方向缩放比率

返回：void
**/

```cpp
void ZoomImage(float fXZoomRatio, float fYZoomRatio)
{
    //判断图像数据类型
    switch (datatype)
    {
    case GDT_Byte:
        ZoomImage(byte(0), fXZoomRatio, fYZoomRatio);
        break;
    case GDT_UInt16:
        ZoomImage(WORD(0), fXZoomRatio, fYZoomRatio);
        break;
    case GDT_Int16:
        ZoomImage(short(0), fXZoomRatio, fYZoomRatio);
        break;
    case GDT_UInt32:
        ZoomImage(DWORD(0), fXZoomRatio, fYZoomRatio);
        break;
    case GDT_Int32:
        ZoomImage(int(0), fXZoomRatio, fYZoomRatio);
        break;
    case GDT_Float32:
        ZoomImage(float(0), fXZoomRatio, fYZoomRatio);
        break;
    case GDT_Float64:
        ZoomImage(double(0), fXZoomRatio, fYZoomRatio);
        break;
    default:
        break;
    }
}
template <class T>
void ZoomImage (T, float fXZoomRatio, float fYZoomRatio)
{
    //判断图像是否为空
    if (! IsValid())
    {
```

```cpp
        return;
}
//获取图像数据指针
T * lpData = (T *)pData;
T * lpDataNew;
int i, j, k;

//源图像的宽度和高度
LONG    lWidth;
LONG    lHeight;

//缩放后图像的宽度和高度
LONG    lNewWidth;
LONG    lNewHeight;

// 指向源像素的指针
T * lpSrc;

// 指向缩放图像对应像素的指针
T * lpDst;

//像素在源图像中的坐标
LONG    i0;
LONG    j0;


//获取图像的宽度
lWidth = GetWidth();
//获取图像的高度
lHeight = GetHeight();

//计算缩放后的图像实际宽度
//此处直接加0.5是由于强制类型转换时不是四舍五入,而是直接截去小数部分
lNewWidth = (LONG)(lWidth * fXZoomRatio + 0.5);

//计算缩放后的图像高度
lNewHeight = (LONG)(lHeight * fYZoomRatio + 0.5);
```

```cpp
// 分配内存，以保存新图像
lpDataNew = new T[lNewWidth * lNewHeight * nBands];

//判断是否内存分配失败
if (lpDataNew == NULL)    return;

//逐波段进行处理
for (k=0; k<nBands; k++)
{
    // 针对图像每行进行操作
    for(i = 0; i < lNewHeight; i++)
    {
        // 针对图像每列进行操作
        for(j = 0; j < lNewWidth; j++)
        {
            // 指向新图像第 i 行，第 j 个像素的指针
            // 注意此处宽度和高度是新图像的宽度和高度
            lpDst = lpDataNew + lNewWidth * i + j +k * lNewWidth * lNewHeight;
            // 计算该像素在源图像中的坐标
            i0 = (LONG) (i / fYZoomRatio + 0.5);
            j0 = (LONG) (j / fXZoomRatio + 0.5);

            // 判断是否在源图范围内
            if((j0 >= 0) && (j0 < lWidth) && (i0 >= 0) && (i0 < lHeight))
            {
                // 指向源图像第 i0 行，第 j0 个像素的指针
                lpSrc = lpData + lWidth * i0 + j0 + k * lWidth * lHeight;
                // 复制像素
                *lpDst = *lpSrc;
            }
            else
            {
                // 对于源图中没有的像素，直接赋值为 0
                *(lpDst) = 0;
            }
        }
    }
```

}
//修改缩放后的宽度与高度,重新创建新位图
nWidth = lNewWidth; nHeight = lNewHeight;
CreateDIB( );
//修改显示数据,更新视图
BYTE *pByte = m_pDIBs;
for (k=0; k<nBandsShow; k++)
{
    //统计原图像最大和最小灰度级
    float m_MinGray = (float)999999999, m_MaxGray = (float)-99999999;
    if (datatype == GDT_Byte)
    {
        m_MaxGray = 255;
        m_MinGray = 0;
    }
    else
    {
        for (i=0; i<nHeight; i++)
        {
            for (j=0; j<nWidth; j++)
            {
                if(*(lpDataNew+nWidth*i+j+k*nWidth*nHeight)<m_MinGray)
                    m_MinGray = *(lpDataNew+nWidth*i+j+k*nWidth*nHeight);
                if(*(lpDataNew+nWidth*i+j+k*nWidth*nHeight)>m_MaxGray)
                    m_MaxGray = *(lpDataNew+nWidth*i+j+k*nWidth*nHeight);
            }
        }
    }
    for (i=0; i<nHeight; i++)
    {
        for (j=0; j<nWidth; j++)
        {
            if (nBandsShow == 1)
(pByte+lLineBYTES(nHeight-1-i)+j) = (BYTE)((*(lpDataNew+nWidth*i+j+k*nWidth*nHeight)-m_MinGray)*255/(m_MaxGray-m_MinGray));
            else
(pByte+lLineBYTES(nHeight-1-i)+j*3+2-k) = (BYTE)((*(lpDataNew+nWidth*i+j+k*nWidth*nHeight)-m_MinGray)*255/(m_MaxGray-m_MinGray));

            }
        }
    }
    delete [ ]lpDataNew;
}

9. 图像旋转

/**

图像旋转

参数：

    int iRotateAngle　旋转的角度(0~360°)

返回： void

**/

void RotateImg(int iRotateAngle)
{
    //判断图像数据类型
    switch (datatype)
    {
    case GDT_Byte：
        RotateImg (byte(0), iRotateAngle)；
        break；
    case GDT_UInt16：
        RotateImg (WORD(0), iRotateAngle)；
        break；
    case GDT_Int16：
        RotateImg (short(0), iRotateAngle)；
        break；
    case GDT_UInt32：
        RotateImg (DWORD(0), iRotateAngle)；
        break；
    case GDT_Int32：
        RotateImg (int(0), iRotateAngle)；
        break；
    case GDT_Float32：
        RotateImg (float(0), iRotateAngle)；
        break；
    case GDT_Float64：
        RotateImg (double(0), iRotateAngle)；

```cpp
            break;
        default:
            break;
        }
}
template <class T>
void RotateImg (T, int iRotateAngle)
{
    //判断图像是否为空
    if (! IsValid())
    {
        return;
    }
    //获取图像数据指针
    T *lpData = (T*)pData;
    T *lpDataNew;
    int i, j, k;

    //源图像的宽度和高度
    LONG    lWidth;
    LONG    lHeight;

    //旋转后图像的宽度和高度
    LONG    lNewWidth;
    LONG    lNewHeight;



    //指向源像素的指针
    T    *lpSrc;

    //指向缩放图像对应像素的指针
    T    *lpDst;

    //像素在源图像中的坐标
    LONG    i0;
    LONG    j0;
```

```cpp
//旋转角度(弧度)
floatfRotateAngle;

//旋转角度的正弦和余弦
floatfSina, fCosa;

//源图四个角的坐标(以图像中心为坐标系原点)
floatfSrcX1, fSrcY1, fSrcX2, fSrcY2, fSrcX3, fSrcY3, fSrcX4, fSrcY4;

//旋转后四个角的坐标(以图像中心为坐标系原点)
floatfDstX1, fDstY1, fDstX2, fDstY2, fDstX3, fDstY3, fDstX4, fDstY4;

//两个中间常量
floatf1, f2;

//获取图像的宽度
lWidth = GetWidth();
//获取图像的高度
lHeight = GetHeight();

//将旋转角度从度转换到弧度
fRotateAngle = ((float)iRotateAngle) * 3.1415926/180;

//计算旋转角度的正弦
fSina = (float) sin((double)fRotateAngle);

//计算旋转角度的余弦
fCosa = (float) cos((double)fRotateAngle);

//计算原图的四个角的坐标(以图像中心为坐标系原点)
fSrcX1 = (float) (- (lWidth-1) / 2);
fSrcY1 = (float) (   (lHeight-1) / 2);
fSrcX2 = (float) (   (lWidth-1) / 2);
fSrcY2 = (float) (   (lHeight-1) / 2);
fSrcX3 = (float) (- (lWidth-1) / 2);
fSrcY3 = (float) (- (lHeight-1) / 2);
fSrcX4 = (float) (   (lWidth-1) / 2);
fSrcY4 = (float) (- (lHeight-1) / 2);
```

//计算新图四个角的坐标(以图像中心为坐标系原点)
```
fDstX1 =  fCosa * fSrcX1 + fSina * fSrcY1;
fDstY1 = -fSina * fSrcX1 + fCosa * fSrcY1;
fDstX2 =  fCosa * fSrcX2 + fSina * fSrcY2;
fDstY2 = -fSina * fSrcX2 + fCosa * fSrcY2;
fDstX3 =  fCosa * fSrcX3 + fSina * fSrcY3;
fDstY3 = -fSina * fSrcX3 + fCosa * fSrcY3;
fDstX4 =  fCosa * fSrcX4 + fSina * fSrcY4;
fDstY4 = -fSina * fSrcX4 + fCosa * fSrcY4;
```

//计算旋转后的图像实际宽度
```
lNewWidth = (LONG) ( max( fabs(fDstX4 - fDstX1), fabs(fDstX3 - fDstX2)) + 0.5);
```

//计算旋转后的图像高度
```
lNewHeight = (LONG) ( max( fabs(fDstY4 - fDstY1), fabs(fDstY3 - fDstY2)) + 0.5);
```

//两个常数,这样不用以后每次都计算了
```
f1 = (float) (-0.5 * (lNewWidth-1) * fCosa - 0.5 * (lNewHeight-1) * fSina
    + 0.5 * (lWidth   -1));
f2 = (float) (0.5 * (lNewWidth-1) * fSina - 0.5 * (lNewHeight-1) * fCosa
    + 0.5 * (lHeight-1));
```

//分配内存,以保存新图像
```
lpDataNew = new T[lNewWidth * lNewHeight * nBands];
```

//判断是否内存分配失败
```
if (lpDataNew == NULL)    return;
```

//逐波段进行处理
```
for (k=0; k<nBands; k++)
{
    // 针对图像每行进行操作
    for(i = 0; i < lNewHeight; i++)
    {
        // 针对图像每列进行操作
        for(j = 0; j < lNewWidth; j++)
        {
```

```cpp
            // 指向新图像第i行，第j个像素的指针
            // 注意此处宽度和高度是新图像的宽度和高度
            lpDst = lpDataNew + lNewWidth * i + j +k * lNewWidth * lNewHeight;

            // 计算该像素在源图像中的坐标
            i0 = (LONG)(-((float)j) * fSina + ((float)i) * fCosa + f2 + 0.5);
            j0 = (LONG)(((float)j) * fCosa + ((float)i) * fSina + f1 + 0.5);

            // 判断是否在源图范围内
            if((j0 >= 0) && (j0 < lWidth) && (i0 >= 0) && (i0 < lHeight))
            {
                // 指向源图像第i0行，第j0个像素的指针
                lpSrc = lpData + lWidth * i0 + j0 + k * lWidth * lHeight;
                // 复制像素
                *lpDst = *lpSrc;
            }
            else
            {
                // 对于源图中没有的像素，直接赋值为0
                *lpDst = 0;
            }
        }
    }
}
//修改旋转后的宽度与高度，重新创建新位图
nWidth = lNewWidth;    nHeight = lNewHeight;
CreateDIB();
//修改显示数据，更新视图
BYTE *pByte = m_pDIBs;
for(k=0; k<nBandsShow; k++)
{
    //统计原图像最大和最小灰度级
    float m_MinGray=(float)999999999, m_MaxGray=(float)-99999999;
    if(datatype == GDT_Byte)
    {
        m_MaxGray = 255;
        m_MinGray = 0;
    }
```

```
            else
            {
                for (i=0; i<nHeight; i++)
                {
                    for (j=0; j<nWidth; j++)
                    {
                        if(*(lpDataNew+nWidth*i+j+k*nWidth*nHeight)<m_MinGray)
                            m_MinGray=*(lpDataNew+nWidth*i+j+k*nWidth*nHeight);
                        if(*(lpDataNew+nWidth*i+j+k*nWidth*nHeight)>m_MaxGray)
                            m_MaxGray=*(lpDataNew+nWidth*i+j+k*nWidth*nHeight);
                    }
                }
            }
            for (i=0; i<nHeight; i++)
            {
                for (j=0; j<nWidth; j++)
                {
                    if (nBandsShow == 1)
            *(pByte+lLineBYTES*(nHeight-1-i)+j)=(BYTE)((*(lpDataNew+nWidth*i+j+k
*nWidth*nHeight)-m_MinGray)*255/(m_MaxGray-m_MinGray));
                    else
            *(pByte+lLineBYTES*(nHeight-1-i)+j*3+2-k)=(BYTE)((*(lpDataNew+nWidth
*i+j+k*nWidth*nHeight)-m_MinGray)*255/(m_MaxGray-m_MinGray));
                }
            }
        }
        delete []lpDataNew;
}
```

10. 图像转置

```
/*************************************************************
图像转置
    参数：          无
    返回：          void
*************************************************************/
void TransposeImg()
{
    //判断图像数据类型
    switch (datatype)
```

```cpp
    }
        case GDT_Byte:
            TransposeImg(byte(0));
            break;
        case GDT_UInt16:
            TransposeImg(WORD(0));
            break;
        case GDT_Int16:
            TransposeImg(short(0));
            break;
        case GDT_UInt32:
            TransposeImg(DWORD(0));
            break;
        case GDT_Int32:
            TransposeImg(int(0));
            break;
        case GDT_Float32:
            TransposeImg(float(0));
            break;
        case GDT_Float64:
            TransposeImg(double(0));
            break;
        default:
            break;
    }
}
template <class T>
void TransposeImg(T)
{
    //判断图像是否为空
    if(! IsValid())
    {
        return;
    }
    //获取图像数据指针
    T *lpData = (T *)pData;
    T *lpDataNew;
    int i, j, k;
```

```
//源图像的宽度和高度
LONG    lWidth;
LONG    lHeight;

//指向源像素的指针
T   * lpSrc;

//指向缩放图像对应像素的指针
T   * lpDst;

//获取图像的宽度
lWidth = GetWidth( );
//获取图像的高度
lHeight = GetHeight( );

//分配内存，以保存新图像
lpDataNew = new T[lWidth * lHeight * nBands];

//判断是否内存分配失败
if (lpDataNew = = NULL)     return;
//逐波段进行处理
for (k=0; k<nBands; k++)
{
    // 针对图像每行进行操作
    for(i = 0; i < lHeight; i++)
    {
        // 针对每行图像每列进行操作
        for(j = 0; j < lWidth; j++)
        {
            // 指向源图像第 i 行，第 j 个像素的指针
            lpSrc = lpData + lWidth * i + j +k * lWidth * lHeight;

            // 指向转置图像第 j 行，第 i 个像素的指针
            // 注意此处 lWidth 和 lHeight 是源图像的宽度和高度，应该互换
            lpDst = lpDataNew + lHeight * (lWidth-1-j) + i+k * lWidth * lHeight;

            // 复制像素
```

```
                    *lpDst = *lpSrc;
            }
        }
    }
    //修改转置后的宽度与高度，重新创建新位图
    nWidth = lHeight;    nHeight = lWidth;
    CreateDIB();
    //修改显示数据，更新视图
    BYTE *pByte = m_pDIBs;
    for (k=0; k<nBandsShow; k++)
    {
        //统计原图像最大和最小灰度级
        float m_MinGray=(float)999999999, m_MaxGray=(float)-99999999;
        if (datatype == GDT_Byte)
        {
            m_MaxGray = 255;
            m_MinGray = 0;
        }
        else
        {
            for (i=0; i<nHeight; i++)
            {
                for (j=0; j<nWidth; j++)
                {
                    if(*(lpDataNew+nWidth*i+j+k*nWidth*nHeight)<m_MinGray)
                        m_MinGray=*(lpDataNew+nWidth*i+j+k*nWidth*nHeight);
                    if(*(lpDataNew+nWidth*i+j+k*nWidth*nHeight)>m_MaxGray)
                        m_MaxGray=*(lpDataNew+nWidth*i+j+k*nWidth*nHeight);
                }
            }
        }
        for (i=0; i<nHeight; i++)
        {
            for (j=0; j<nWidth; j++)
            {
                if (nBandsShow == 1)
                    *(pByte+lLineBYTES*(nHeight-1-i)+j)=(BYTE)((*(lpDataNew+nWidth*i+j+k*nWidth*nHeight)-m_MinGray)*255/(m_MaxGray-m_MinGray));
```

```
                else
    *(pByte+lLineBYTES*(nHeight-1-i)+j*3+2-k)=(BYTE)((*(lpDataNew+nWidth
*i+j+k*nWidth*nHeight)-m_MinGray)*255/(m_MaxGray-m_MinGray));
            }
        }
    }
    delete []lpDataNew;
}
```

11. 二维傅立叶变换

```
/*************************************************************
二维傅立叶变换
    参数：
    complex<double> * pCTData    图像数据
    int      nWidth              数据宽度
    int      nHeight             数据高度
    complex<double> * pCFData    傅立叶变换后的结果

    返回：   无
*************************************************************/
void Fourie(complex<double> * pCTData, int nWidth, int nHeight, complex<double>
* pCFData)
{
    //循环控制变量
    int  x;
    int  y;

    //临时变量
    double   dTmpOne;
    double   dTmpTwo;

    //进行傅立叶变换的宽度和高度(2的整数次幂)
    //图像的宽度和高度不一定为2的整数次幂
    int      nTransWidth;
    int      nTransHeight;

    //计算进行傅立叶变换的宽度(2的整数次幂)
    dTmpOne = log(nWidth)/log(2);
    dTmpTwo = ceil(dTmpOne)            ;
```

```cpp
        dTmpTwo = pow(2, dTmpTwo)      ;
        nTransWidth = (int) dTmpTwo    ;

        //计算进行傅立叶变换的高度 (2 的整数次幂)
        dTmpOne = log(nHeight)/log(2);
        dTmpTwo = ceil(dTmpOne)        ;
        dTmpTwo = pow(2, dTmpTwo)      ;
        nTransHeight = (int) dTmpTwo   ;

        // x, y(行列)方向上的迭代次数
        int     nXLev;
        int     nYLev;

        //计算 x, y(行列)方向上的迭代次数
        nXLev = (int) ( log(nTransWidth)/log(2) +0.5 );
        nYLev = (int) ( log(nTransHeight)/log(2) + 0.5 );

        for(y = 0; y < nTransHeight; y++){
            // x 方向进行快速傅立叶变换
            FastFourie(&pCTData[nTransWidth * y], &pCFData[nTransWidth * y], nXLev);
        }

        // pCFData 中目前存储了 pCTData 经过行变换的结果
        //为了直接利用 FFT_1D,需要把 pCFData 的二维数据转置,再一次利用 FFT_1D 进行
        //傅立叶行变换(实际上相当于对列进行傅立叶变换)
        for(y = 0; y < nTransHeight; y++){
            for(x = 0; x < nTransWidth; x++){
                pCTData[nTransHeight * x + y] = pCFData[nTransWidth * y + x];
            }}

        for(x = 0; x < nTransWidth; x++){
            // 对 x 方向进行快速傅立叶变换,实际上相当于对原来的图像数据进行列方向的
            // 傅立叶变换
            FastFourie(&pCTData[x * nTransHeight], &pCFData[x * nTransHeight], nYLev);
        }
```

// pCFData 中目前存储了 pCTData 经过二维傅立叶变换的结果,但是为了方便列方向
//的傅立叶变换,对其进行了转置,现在把结果转置回来
    for(y = 0; y < nTransHeight; y++){
        for(x = 0; x < nTransWidth; x++){
            pCTData[nTransWidth * y + x] = pCFData[nTransHeight * x + y];
        }}

    memcpy(pCTData, pCFData, sizeof(complex<double>) * nTransHeight * nTransWidth);
}
/***
快速傅立叶变换
参数:
    complex<double> * pCTData   指向时域数据的指针,输入的需要变换的数据
    complex<double> * pCFData   指向频域数据的指针,输出的经过变换的数据
    nLevel   傅立叶变换蝶形算法的级数,2 的幂数,

返回:
***/
void FastFourie(complex<double> * pCTData, complex<double> * pCFData, int nLevel)
{
    //循环控制变量
    int   i;
    int   j;
    int   k;

    double PI = 3.1415926;

    //傅立叶变换点数
    int nCount =0 ;

    //计算傅立叶变换点数
    nCount =(int)pow(2, nLevel) ;

    //某一级的长度
    int   nBtFlyLen;
    nBtFlyLen = 0 ;

    //变换系数的角度 =2 * PI * i / nCount

```cpp
            double    dAngle;

complex<double>  * pCW ;

//分配内存,存储傅立叶变化需要的系数表
pCW = new complex<double>[ nCount / 2 ];

//计算傅立叶变换的系数
for( i = 0; i < nCount / 2; i++){
    dAngle = -2 * PI * i / nCount;
    pCW[ i ] = complex<double> ( cos( dAngle ), sin( dAngle ));
}

//变换需要的工作空间
complex<double>  * pCWork1 ,  * pCWork2;

//分配工作空间
pCWork1 = new complex<double>[ nCount ];
pCWork2 = new complex<double>[ nCount ];

//临时变量
complex<double>  * pCTmp;

//初始化,写入数据
memcpy( pCWork1, pCTData, sizeof( complex<double>) * nCount );

//临时变量
int nInter;
nInter = 0;

//蝶形算法进行快速傅立叶变换
for( k = 0; k < nLevel; k++){
    for( j = 0; j < (int)pow(2, k); j++){
        //计算长度
        nBtFlyLen = (int)pow( 2, (nLevel-k));

        //倒序重排,加权计算
        for( i = 0; i < nBtFlyLen/2; i++){
```

```
                nInter = j * nBtFlyLen;
                pCWork2[i + nInter] =
                    pCWork1[i + nInter] + pCWork1[i + nInter + nBtFlyLen / 2];
                pCWork2[i + nInter + nBtFlyLen / 2] =
                    (pCWork1[i + nInter] - pCWork1[i + nInter + nBtFlyLen / 2])
                        * pCW[(int)(i * pow(2, k))];
            }}

    // 交换 pCWork1 和 pCWork2 的数据
    pCTmp   = pCWork1;
    pCWork1 = pCWork2;
    pCWork2 = pCTmp;
}

//重新排序
for(j = 0; j < nCount; j++){
    nInter = 0;
    for(i = 0; i < nLevel; i++){
        if ( j&(1<<i))    nInter += 1<<(nLevel-i-1);
    }
    pCFData[j]=pCWork1[nInter];
}

//释放内存空间
delete pCW;
delete pCWork1;
delete pCWork2;
pCW     = NULL;
pCWork1 = NULL;
pCWork2 = NULL;
}
```

12. 二维傅立叶反变换

```
/***************************************************************
二维傅立叶反变换
参数：
    complex<double> * pCFData    频域数据
    complex<double> * pCTData    时域数据
    int      nWidth              图像数据宽度
```

```
    int    nHeight                图像数据高度

返回: 无
*************************************************************/
void IFourie( complex<double> * pCFData, complex<double> * pCTData, int nWidth, int nHeight)
{
    //循环控制变量
    int x;
    int y;

    //临时变量
    double   dTmpOne;
    double   dTmpTwo;

    //进行傅立叶变换的宽度和高度, (2 的整数次幂)
    //图像的宽度和高度不一定为 2 的整数次幂
    int    nTransWidth;
    int  nTransHeight;

    //计算进行傅立叶变换的宽度(2 的整数次幂)
    dTmpOne = log(nWidth)/log(2);
    dTmpTwo = ceil(dTmpOne)           ;
    dTmpTwo = pow(2, dTmpTwo)          ;
    nTransWidth = (int) dTmpTwo        ;

    //计算进行傅立叶变换的高度 (2 的整数次幂)
    dTmpOne = log(nHeight)/log(2);
    dTmpTwo = ceil(dTmpOne)           ;
    dTmpTwo = pow(2, dTmpTwo)          ;
    nTransHeight = (int) dTmpTwo       ;

    //分配工作需要的内存空间
    complex<double> * pCWork= new complex<double>[nTransWidth * nTransHeight];

    //临时变量
    complex<double> * pCTmp ;
```

```
        //为了利用傅立叶正变换,可以把傅立叶频域的数据取共轭
        //然后直接利用正变换,输出结果就是傅立叶反变换结果的共轭
        for( y = 0; y < nTransHeight; y++){
            for( x = 0; x < nTransWidth; x++){
                pCTmp = &pCFData[nTransWidth * y+x];
                pCWork[nTransWidth * y+x] = complex<double>( pCTmp->real( ), -pCTmp->imag( ));
            }
        }

        //调用傅立叶正变换
        Fourie(pCWork, nWidth, nHeight, pCTData);
        // 求时域点的共轭,求得最终结果
        //根据傅立叶变换原理,利用这样的方法求得的结果和实际的时域数据
        //相差一个系数
        for( y = 0; y < nTransHeight; y++){
            for( x = 0; x < nTransWidth; x++){
                pCTmp = &pCTData[nTransWidth * y + x];
                pCTData[nTransWidth * y + x] =
                    complex<double>( pCTmp->real( )/(nTransWidth * nTransHeight),
                    -pCTmp->imag( )/(nTransWidth * nTransHeight));
            }
        }
        delete pCWork;
        pCWork = NULL;
}
```

13. Butterworth 低通滤波

/ ***

ButterWorth 低通滤波
参数:
    LPBYTE lpImage   指向需要增强得图像数据
    int nWidth   数据宽度
    int nHeight   数据高度
    int nRadius   ButterWorth 低通滤波的"半功率"点

返回: 无
***/
void ButterWorthLowPass(LPBYTE lpImage, int nWidth, int nHeight, int nRadius)
{
    //循环控制变量

```cpp
    int y ;
    int x ;

    double dTmpOne ;
    double dTmpTwo ;

    // ButterWorth 滤波系数
    double H         ;

    //傅立叶变换的宽度和高度(2 的整数次幂)
    int nTransWidth ;
    int nTransHeight;

    double dReal     ;
    double dImag     ;
    //图像像素值
    unsigned char unchValue;

    //指向时域数据的指针
    complex<double> * pCTData;
    //指向频域数据的指针
    complex<double> * pCFData ;

    //计算进行傅立叶变换的点数(2 的整数次幂)
    dTmpOne = log(nWidth)/log(2);
    dTmpTwo = ceil(dTmpOne)      ;
    dTmpTwo = pow(2, dTmpTwo)    ;
    nTransWidth = (int) dTmpTwo  ;

    //计算进行傅立叶变换的点数 (2 的整数次幂)
    dTmpOne = log(nHeight)/log(2);
    dTmpTwo = ceil(dTmpOne)      ;
    dTmpTwo = pow(2, dTmpTwo)    ;
    nTransHeight = (int) dTmpTwo ;

    //分配内存
    pCTData=new complex<double>[nTransWidth * nTransHeight];
    pCFData=new complex<double>[nTransWidth * nTransHeight];
```

//初始化
//图像数据的宽和高不一定是2的整数次幂，所以pCTData
//有一部分数据需要补0
for(y=0; y<nTransHeight; y++){
    for(x=0; x<nTransWidth; x++){
        pCTData[y*nTransWidth + x]=complex<double>(0, 0);
    }
}

//把图像数据传给pCTData
for(y=0; y<nHeight; y++){
    for(x=0; x<nWidth; x++){
        unchValue = lpImage[y*nWidth +x];
        pCTData[y*nTransWidth + x]=complex<double>(unchValue, 0);
    }
}

//傅立叶正变换
Fourie(pCTData, nWidth, nHeight, pCFData);

//下面开始实施ButterWorth低通滤波
for(y=0; y<nTransHeight; y++){
    for(x=0; x<nTransWidth; x++){
        H = (double)(y*y+x*x);
        H = H / (nRadius * nRadius);
        H = 1/(1+H)   ;
        pCFData[y*nTransWidth + x]=complex<double>(pCFData[y*nTransWidth + x].real()*H,
            pCFData[y*nTransWidth + x].imag()*H);
    }
}

//经过ButterWorth低通滤波的图像进行反变换
IFourie(pCFData, pCTData, nWidth, nHeight);

//反变换的数据传给lpImage
for(y=0; y<nHeight; y++){
    for(x=0; x<nWidth; x++){
        dReal = pCTData[y*nTransWidth + x].real();
        dImag = pCTData[y*nTransWidth + x].imag();

```
                unchValue = (unsigned char)max(0, min(255, sqrt(dReal * dReal+dImag *
dImag)));
                lpImage[y * nWidth + x] = unchValue;
        }}
```

//释放内存
```
delete pCTData;
delete pCFData;
pCTData = NULL;
pCFData = NULL;
}
```

14. Butterworth 高通滤波

/***

ButterWorth 高通滤波

参数：

  LPBYTE lpImage   指向需要增强得图像数据

  int nWidth   数据宽度

  int nHeight   数据高度

  int nRadius   ButterWorth 高通滤波的"半功率"点

返回： 无

***/

```
void ButterWorthHighPass(LPBYTE lpImage, int nWidth, int nHeight, int nRadius)
{
    //循环控制变量
    int y ;
    int x ;

    double dTmpOne ;
    double dTmpTwo ;

    // ButterWorth 滤波系数
    double H           ;

    //傅立叶变换的宽度和高度(2 的整数次幂)
    int nTransWidth ;
    int nTransHeight;
```

```cpp
        double dReal    ;
        double dImag    ;
//图像像素值
unsigned char unchValue;

//指向时域数据的指针
complex<double> * pCTData;
//指向频域数据的指针
complex<double> * pCFData;

//计算进行傅立叶变换的点数(2 的整数次幂)
dTmpOne = log(nWidth)/log(2);
dTmpTwo = ceil(dTmpOne)    ;
dTmpTwo = pow(2, dTmpTwo)   ;
nTransWidth = (int) dTmpTwo ;

//计算进行傅立叶变换的点数 (2 的整数次幂)
dTmpOne = log(nHeight)/log(2);
dTmpTwo = ceil(dTmpOne)    ;
dTmpTwo = pow(2, dTmpTwo)   ;
nTransHeight = (int) dTmpTwo ;

//分配内存
pCTData=new complex<double>[nTransWidth * nTransHeight];
pCFData=new complex<double>[nTransWidth * nTransHeight];

//初始化
//图像数据的宽和高不一定是 2 的整数次幂,所以 pCTData
//有一部分数据需要补 0
for(y=0; y<nTransHeight; y++) {
    for(x=0; x<nTransWidth; x++) {
        pCTData[y * nTransWidth + x]=complex<double>(0, 0);
    }}

//把图像数据传给 pCTData
for(y=0; y<nHeight; y++) {
    for(x=0; x<nWidth; x++) {
        unchValue = lpImage[y * nWidth +x];
```

```cpp
            pCTData[y * nTransWidth + x] = complex<double>(unchValue, 0);
        }
    }

    //傅立叶正变换
    Fourie(pCTData, nWidth, nHeight, pCFData);

    //下面开始实施 ButterWorth 高通滤波
    for(y=0; y<nTransHeight; y++) {
        for(x=0; x<nTransWidth; x++) {

            H = (double)(y*y+x*x);
            H = (nRadius * nRadius) / H;
            H = 1/(1+H)            ;
            pCFData[y * nTransWidth + x] = complex<double>(H * (pCFData[y * nTransWidth + x].real()),
                    H * (pCFData[y * nTransWidth + x].imag()) );
        }
    }

    //经过 ButterWorth 高通滤波的图像进行反变换
    IFourie(pCFData, pCTData, nWidth, nHeight);

    //反变换的数据传给 lpImage
    for(y=0; y<nHeight; y++) {
        for(x=0; x<nWidth; x++) {
            dReal = pCTData[y * nTransWidth + x].real();
            dImag = pCTData[y * nTransWidth + x].imag();
            unchValue = (unsigned char)max(0, min(255, sqrt(dReal * dReal+dImag * dImag)+100));
            lpImage[y * nWidth + x] = unchValue;
        }
    }

    //释放内存
    delete pCTData;
    delete pCFData;
    pCTData = NULL;
    pCFData = NULL;
}
```

15. 边缘检测

```
/****************************************************
Roberts 算子
参数 r：         void
返回：           BOOL
****************************************************/
BOOL RobertsImg( )
    {
        //判断图像数据格式类型
        switch（datatype）
        {
        case GDT_Byte：
            RobertsImg( byte(0) ) ;
            break ;
        case GDT_UInt16：
            RobertsImg ( WORD(0) ) ;
            break ;
        case GDT_Int16：
            RobertsImg ( short(0) ) ;
            break ;
        case GDT_UInt32：
            RobertsImg ( DWORD(0) ) ;
            break ;
        case GDT_Int32：
            RobertsImg ( int(0) ) ;
            break ;
        case GDT_Float32：
            RobertsImg ( float(0) ) ;
            break ;
        case GDT_Float64：
            RobertsImg ( double(0) ) ;
            break ;
        default：
            break ;
        }
        return TRUE ;
    }
template <class T>
```

```cpp
BOOL RobertsImg(T)
{
    //判断图像是否为空
    if(! IsValid())
    {
        return FALSE;
    }
    //获取图像数据指针
    T *lpData = (T *)pData;
    //获取图像的宽度与高度
    DWORD nWidth = GetWidth();
    DWORD nHeight = GetHeight();
    //分配内存
    T *lpDataCopy = new T[nHeight*nWidth*nBands];
    //获取显示数据指针
    BYTE *pByte = m_pDIBs;
    //拷贝源数据
    memcpy(lpDataCopy, lpData, nWidth*nHeight*nBands*sizeof(T));
    //图像变换开始
    int i, j, k;
    //逐波段进行处理
    for(k=0; k<nBands; k++)
    {
        for(i=0; i<nHeight-1; i++)
            for(j=0; j<nWidth-1; j++)
            {
                *(lpData+i*nWidth+j+k*nWidth*nHeight) =
                    abs(*(lpDataCopy+(i+1)*nWidth+j+1+k*nWidth*nHeight)
                    - *(lpDataCopy+i*nWidth+j+k*nWidth*nHeight));
            }
    }
    ///修改显示数据,更新视图
    for(k=0; k<nBandsShow; k++)
    {
        //统计原图像最大和最小灰度级
        float m_MinGray=(float)999999999, m_MaxGray=(float)-99999999;
        if(datatype == GDT_Byte)
        {
```

```
                m_MaxGray = 255;
                m_MinGray = 0;
            }
            else
            {
                for(i=0; i<nHeight; i++)
                {
                    for(j=0; j<nWidth; j++)
                    {
                        if(*(lpData+nWidth*i+j+k*nWidth*nHeight)<m_MinGray)
                            m_MinGray = *(lpData+nWidth*i+j+k*nWidth*nHeight);
                        if(*(lpData+nWidth*i+j+k*nWidth*nHeight)>m_MaxGray)
                            m_MaxGray = *(lpData+nWidth*i+j+k*nWidth*nHeight);
                    }
                }
            }
            for(i=0; i<nHeight; i++)
            {
                for(j=0; j<nWidth; j++)
                {
                    if(nBandsShow == 1)
    *(pByte+lLineBYTES*(nHeight-1-i)+j) = (BYTE)((*(lpData+nWidth*i+j+k*
nWidth*nHeight)-m_MinGray)*255/(m_MaxGray-m_MinGray));
                    else
    *(pByte+lLineBYTES*(nHeight-1-i)+j*3+2-k) = (BYTE)((*(lpData+nWidth*i
+j+k*nWidth*nHeight)-m_MinGray)*255/(m_MaxGray-m_MinGray));
                }
            }
        }
        delete []lpDataCopy;
        return TRUE;
    }
/***************************************************************
Prewitt 算子
参数 r:          int tag
返回:            BOOL
***************************************************************/
    BOOL PrewittImg(int tag)
```

```cpp
    {
        //判断图像数据类型
        switch (datatype)
        {
        case GDT_Byte：
            PrewittImg (byte(0), tag);
            break;
        case GDT_UInt16：
            PrewittImg (WORD(0), tag);
            break;
        case GDT_Int16：
            PrewittImg (short(0), tag);
            break;
        case GDT_UInt32：
            PrewittImg (DWORD(0), tag);
            break;
        case GDT_Int32：
            PrewittImg (int(0), tag);
            break;
        case GDT_Float32：
            PrewittImg (float(0), tag);
            break;
        case GDT_Float64：
            PrewittImg (double(0), tag);
            break;
        default：
            break;
        }
        return TRUE;
    }
    template <class T>
    BOOL PrewittImg (T, int tag)
    {
        //判断图像是否为空
        if (! IsValid( ))
        {
            return FALSE;
        }
```

//获取图像数据指针
T *lpData = (T *)pData;
//获取图像的宽度与高度
DWORD nWidth = GetWidth();
DWORD nHeight = GetHeight();
//分配内存
T *lpDataCopy = new T[nHeight * nWidth * nBands];
//获取显示数据指针
BYTE *pByte = m_pDIBs;
//拷贝源数据
memcpy(lpDataCopy, lpData, nWidth * nHeight * nBands * sizeof(T));

//图像变换开始
int i, j, k, m, n;
float sum;
int t, Prewitt[9];

//设置 Prewitt 算子
if(tag==0)//检测垂直
{
    Prewitt[0] = -1;
    Prewitt[1] = 0;
    Prewitt[2] = 1;
    Prewitt[3] = -1;
    Prewitt[4] = 0;
    Prewitt[5] = 1;
    Prewitt[6] = -1;
    Prewitt[7] = 0;
    Prewitt[8] = 1;
}
else if(tag==1)//检测水平
{
    Prewitt[0] = -1;
    Prewitt[1] = -1;
    Prewitt[2] = -1;
    Prewitt[3] = 0;
    Prewitt[4] = 0;
    Prewitt[5] = 0;

```cpp
            Prewitt[6] = 1;
            Prewitt[7] = 1;
            Prewitt[8] = 1;
    }
    else return FALSE;

    //逐波段进行处理
    for (k=0; k<nBands; k++)
    {
        for(i=1; i<nHeight-1; i++)
            for(j=1; j<nWidth-1; j++)
            {
                sum=0; t=0;
                //3X3 模板运算
                for(m=0; m<3; m++)
                {
                    for(n=0; n<3; n++)
                    {
                        sum+= * (lpDataCopy +nWidth * (i-1+m)+(j-1+n)+k * nWidth * nHeight) * Prewitt[t++];
                    }
                }
                if (datatype == GDT_Byte)
                {
                    if(sum>255) * (lpData+nWidth * i+j+k * nWidth * nHeight)= 255;
                    else if(sum<0) * (lpData+nWidth * i+j+k * nWidth * nHeight)= 0;
                    else * (lpData+nWidth * i+j+k * nWidth * nHeight)= (BYTE)sum;
                }
                else
                    * (lpData +nWidth * i+j+k * nWidth * nHeight) = (T)sum;
            }
    }

    //修改显示数据，更新视图
    for (k=0; k<nBandsShow; k++)
    {
        //统计原图像最大和最小灰度级
        float m_MinGray = (float)999999999, m_MaxGray = (float)-99999999;
        if (datatype == GDT_Byte)
```

```
                }
                    m_MaxGray = 255;
                    m_MinGray = 0;
            }
                else
            {
                    for(i=0; i<nHeight; i++)
                {
                        for(j=0; j<nWidth; j++)
                    {
                            if( *(lpData+nWidth*i+j+k*nWidth*nHeight)<m_MinGray)
                                m_MinGray = *(lpData+nWidth*i+j+k*nWidth*nHeight);
                            if( *(lpData+nWidth*i+j+k*nWidth*nHeight)>m_MaxGray)
                                m_MaxGray = *(lpData+nWidth*i+j+k*nWidth*nHeight);
                    }
                }
                    for(i=0; i<nHeight; i++)
                {
                        for(j=0; j<nWidth; j++)
                    {
                            if(nBandsShow == 1)
    *(pByte+lLineBYTES*(nHeight-1-i)+j)=(BYTE)((*(lpData+nWidth*i+j+k*
nWidth*nHeight)-m_MinGray)*255/(m_MaxGray-m_MinGray));
                            else
    *(pByte+lLineBYTES*(nHeight-1-i)+j*3+2-k)=(BYTE)((*(lpData+nWidth*i
+j+k*nWidth*nHeight)-m_MinGray)*255/(m_MaxGray-m_MinGray));
                    }
                }
            }
            delete []lpDataCopy;
            return TRUE;
    }
/***********************************************************
Sobel 算子
参数:       int tag
返回:       BOOL
***********************************************************/
```

```cpp
BOOL SobelImg( int tag)
{
    //判断图像数据类型
    switch (datatype)
    {
    case GDT_Byte:
        SobelImg(byte(0), tag);
        break;
    case GDT_UInt16:
        SobelImg (WORD(0), tag);
        break;
    case GDT_Int16:
        SobelImg (short(0), tag);
        break;
    case GDT_UInt32:
        SobelImg (DWORD(0), tag);
        break;
    case GDT_Int32:
        SobelImg (int(0), tag);
        break;
    case GDT_Float32:
        SobelImg (float(0), tag);
        break;
    case GDT_Float64:
        SobelImg (double(0), tag);
        break;
    default:
        break;
    }
    return TRUE;
}
template <class T>
BOOL SobelImg (T, int tag)
{
    //判断图像是否为空
    if (! IsValid())
    {
        return FALSE;
```

}
//获取图像数据指针
T *lpData = (T *)pData;
//获取图像的宽度与高度
DWORD nWidth = GetWidth();
DWORD nHeight = GetHeight();
//分配内存
T *lpDataCopy = new T[nHeight*nWidth*nBands];
//获取显示数据指针
BYTE *pByte = m_pDIBs;
//拷贝源数据
memcpy(lpDataCopy, lpData, nWidth*nHeight*nBands*sizeof(T));
//图像变换开始
int i, j, k, m, n;
float sum;
int t, Sobel[9];

//设置 Sobel 算子
if(tag==0)//检测垂直
{
    Sobel[0] = -1;
    Sobel[1] = 0;
    Sobel[2] = 1;
    Sobel[3] = -2;
    Sobel[4] = 0;
    Sobel[5] = 2;
    Sobel[6] = -1;
    Sobel[7] = 0;
    Sobel[8] = 1;
}
else if(tag==1)//检测水平
{
    Sobel[0] = -1;
    Sobel[1] = -2;
    Sobel[2] = -1;
    Sobel[3] = 0;
    Sobel[4] = 0;
    Sobel[5] = 0;

```cpp
            Sobel[6] =   1;
            Sobel[7] =   2;
            Sobel[8] =   1;
    }
    else return FALSE;

    //逐波段进行处理
    for (k=0; k<nBands; k++)
    {
        for(i=1; i<nHeight-1; i++)
            for(j=1; j<nWidth-1; j++)
            {
                sum=0; t=0;
                //3X3模板运算
                for(m=0; m<3; m++)
                {
                    for(n=0; n<3; n++)
                    {
                        sum+= * (lpDataCopy +nWidth * (i-1+m)+(j-1+n)+k * nWidth * nHeight) * Sobel[t++];
                    }
                }
                if (datatype == GDT_Byte)
                {
                    if (sum > 255) * (lpData +nWidth * i+j+k * nWidth * nHeight)= 255;
                    else if (sum < 0) * (lpData +nWidth * i+j+k * nWidth * nHeight) = 0;
                    else * (lpData +nWidth * i+j+k * nWidth * nHeight) = (BYTE)sum;
                }
                else
                    * (lpData +nWidth * i+j+k * nWidth * nHeight) = (T)sum;
            }
    }
    //修改显示数据,更新视图
    for (k=0; k<nBandsShow; k++)
    {
        //统计原图像最大和最小灰度级
        float m_MinGray=(float)999999999, m_MaxGray=(float)-99999999;
        if (datatype == GDT_Byte)
```

```
            {
                m_MaxGray = 255;
                m_MinGray = 0;
            }
            else
            {
                for (i=0; i<nHeight; i++)
                {
                    for (j=0; j<nWidth; j++)
                    {
                        if(*(lpData+nWidth*i+j+k*nWidth*nHeight)<m_MinGray)
                            m_MinGray = *(lpData+nWidth*i+j+k*nWidth*nHeight);
                        if(*(lpData+nWidth*i+j+k*nWidth*nHeight)>m_MaxGray)
                            m_MaxGray = *(lpData+nWidth*i+j+k*nWidth*nHeight);
                    }
                }
            }
            for (i=0; i<nHeight; i++)
            {
                for (j=0; j<nWidth; j++)
                {
                    if (nBandsShow == 1)
    *(pByte+lLineBYTES*(nHeight-1-i)+j) = (BYTE)((*(lpData+nWidth*i+j+k*
nWidth*nHeight)-m_MinGray)*255/(m_MaxGray-m_MinGray));
                    else
    *(pByte+lLineBYTES*(nHeight-1-i)+j*3+2-k) = (BYTE)((*(lpData+nWidth*i+j
+k*nWidth*nHeight)-m_MinGray)*255/(m_MaxGray-m_MinGray));
                }
            }
        }
    delete []lpDataCopy;
    return TRUE;
}
```

16. 状态法

/***
状态法(峰谷法)

参数：   void

返回： BOOL
**/

```cpp
BOOL StateMethord( )
{
    if( bitinfor = = NULL )     return FALSE;
    //获取图像的一般信息
    DWORD nWidth = GetWidth( );
    DWORD nHeight = GetHeight( );
    WORD wBitCount = bitinfor->bmiHeader.biBitCount;
    DWORD lRowBytes = WIDTHBYTES( nWidth * ( ( DWORD )wBitCount ) );
    LPBYTE lpData = m_pDIBs;


    //峰谷法获取阈值
    /////////////////////////////////////////
    //定义阈值 nThreshold
    int nThreshold = 0;
    int nNewThreshold = 0;

    //定义并初始化灰度统计数组 hist[256]
    int hist[256];
    memset( hist, 0, sizeof( hist ) );

    // lS1，lS2 分别代表类 1 像素总数、类 2 像素总数
    int lS1, lS2;

    //lP1，lP2 分别表示类 1 质量矩和类 2 质量矩

    double lP1, lP2;

    //meanvalue1，meanvalue2 分别代表类 1 灰度均值和类 2 灰度均值
    double meanvalue1, meanvalue2;

    //迭代次数
    int IterationTimes;

    //灰度最大、最小值
    int graymin = 255, graymax = 0;
```

```cpp
    DWORD i, j;
    int k;
    if(wBitCount == 8){
        //统计出各个灰度值的像素数
        for( i=0; i<nHeight; i++)
            for( j=0; j<nWidth; j++){
                int gray = *(lpData+lRowBytes * i+j);
                hist[gray]++;
                if(gray>graymax)         graymax = gray;
                if(gray<graymin)         graymin = gray;
            }

        //给阈值赋迭代初值
        nNewThreshold = int(((graymax+graymin)/2);

        //迭代求最佳阈值
        for(IterationTimes = 0; nThreshold ! = nNewThreshold && IterationTimes<100; IterationTimes ++){
            nThreshold = nNewThreshold;
            lP1 = 0.0;
            lP2 = 0.0;
            lS1 = 0;
            lS2 = 0;

            //求类1的质量矩和像素总数
            for (k = graymin; k <= nThreshold; k++){
                lP1 += (double) k * hist[k];
                lS1 += hist[k];
            }
            //计算类1均值
            meanvalue1 = lP1 / lS1;

            //求类2的质量矩和像素总数
            for (k = nThreshold+1; k <= graymax; k++){
                lP2 += (double) k * hist[k];
                lS2+= hist[k];
            }
            //计算类1均值
```

```
                meanvalue2 = lP2 / lS2;

            //获取新的阈值
            nNewThreshold =   int((meanvalue1 + meanvalue2)/2);
        }
        CString strInfo;
        strInfo.Format("阈值=%d \ n", nThreshold);
        MessageBox(NULL, strInfo,"阈值结果", MB_OK);
    }
    /////////////////////////////////////////////
    //二值化图像
    for(i=0; i<nHeight; i++)
        for(j=0; j<nWidth; j++){
            if( *(lpData+lRowBytes*i+j)<nThreshold)  *(lpData+lRowBytes*i+j)=0;
            else      *(lpData+lRowBytes*i+j)=255;
        }
        return TRUE;
}
```

17. 判断分析法

```
/*************************************************************
判断分析法
参数：    void
返回：    BOOL
**************************************************************/
BOOL AnalyseMethord()
{
    if(bitinfor == NULL)       return FALSE;
    //获取图像的一般信息
    DWORD nWidth = GetWidth();
    DWORD nHeight = GetHeight();
    WORD wBitCount = bitinfor->bmiHeader.biBitCount;
    DWORD lRowBytes = WIDTHBYTES(nWidth*((DWORD)wBitCount));
    LPBYTE lpData = m_pDIBs;

    //判断分析法(Ostu)获取阈值
    /////////////////////////////////////////////
    //定义阈值 nThreshold
    int nThreshold=0;
```

```cpp
//定义并初始化灰度统计数组 hist[256]
int hist[256];
memset(hist, 0, sizeof(hist));

// lS, lS1, lS2 分别代表像素总数、类 1 像素总数、类 2 像素总数
int lS, lS1, lS2;

//lP, lP1 分别表示总的质量矩和类 1 质量矩
double lP, lP1;

//meanvalue1, meanvalue2 分别代表类 1 灰度均值和类 2 灰度均值
double meanvalue1, meanvalue2;

//Dispersion1, Dispersion2, classinDis, classoutDis 表示类 1 方差, 类 2 方差, 类内方差, 类间方差, max 表示类间最大差距
double   Dispersion1, Dispersion2, classinDis, classoutDis, max;

//灰度最大、最小值
int graymin = 255, graymax = 0;

DWORD i, j;
int k = 0;
if(wBitCount == 8){
    //统计出各个灰度值的像素数
    for( i = 0; i<nHeight; i++)
        for( j = 0; j<nWidth; j++){
            int gray = *(lpData+lRowBytes * i+j);
            hist[gray]++;
            if(gray>graymax)            graymax = gray;
            if(gray<graymin)            graymin = gray;
        }

    lP = 0.0;
    lS = 0;
    if(graymin = = 0)    graymin++;

    //计算总的质量矩 lP 和像素总数 lS
```

```cpp
    for( k = graymin; k<= graymax; k++)    {
        lP+=(double)k * (double)hist[k];
        lS+=hist[k];
    }

    max = 0. 0;
    lS1 = 0;
    lS2 = 0;
    lP1 = 0. 0;
    Dispersion1 = 0. 0;
    Dispersion2 = 0. 0;
    classinDis = 0. 0;
    classoutDis = 0. 0;
    double ratio = 0. 0;

    //求阈值
    for( k = graymin; k<= graymax; k++)    {
        //计算类 1 像素总数
        lS1+=hist[k];

        if ( ! lS1)    { continue; }
        //计算类 2 像素总数
        lS2 = lS - lS1;
        if ( lS2 == 0)       { break; }

        //计算类 1 质量矩
        lP1 += (double) k * hist[k];

        //计算类 1 均值
        meanvalue1 = lP1 / lS1;

        //计算类 1 间方差
        for( int n = graymin; n<= k; n++)
            Dispersion1 += ((n-meanvalue1) * (n-meanvalue1) * hist[n]);

        //计算类 2 均值
        meanvalue2 = (lP - lP1) / lS2;
```

```
            //计算类2间方差
            for(int m=k+1; m<=graymax; m++)
                    Dispersion2+=((m-meanvalue2)*(m-meanvalue2)*hist[m]);

        //计算类内方差
            classinDis=Dispersion1+Dispersion2;

        //计算类间方差
             classoutDis = (double)lS1 * (double)lS2 * (meanvalue1 - meanvalue2) *
(meanvalue1 - meanvalue2);

        //类间方差与类内方差比值
            if(classinDis! =0)                    ratio=classoutDis/classinDis;

        //获取分割阈值 nThreshold

            if (ratio> max)    {
               max = ratio;
               nThreshold = k;
            } }

        CString strInfo;
        strInfo. Format("阈值=%d \ n", nThreshold);
        MessageBox(NULL, strInfo,"阈值结果", MB_OK);
    }
 /////////////////////////////////////////////////////
    //二值化图像
     for(i=0; i<nHeight; i++)
            for(j=0; j<nWidth; j++){
                if( *(lpData+lRowBytes*i+j)<nThreshold) *(lpData+lRowBytes*i+j)=0;
                else        *(lpData+lRowBytes*i+j)=255;
            }
        return TRUE;
```

}

18. 最佳熵值法
/ ***
最佳熵值法

参数: void
返回: BOOL
**/
BOOL Shanoon( )
{
  if(bitinfor = = NULL) return FALSE;
  //获取图像的一般信息
  DWORD nWidth = GetWidth( );
  DWORD nHeight = GetHeight( );
  WORD wBitCount = bitinfor->bmiHeader. biBitCount;
  DWORD lRowBytes = WIDTHBYTES(nWidth * ((DWORD)wBitCount));
  LPBYTE lpData = m_ pDIBs;

  //熵值法获取阈值
  /////////////////////////////////
  //定义阈值 nThreshold
  int nThreshold = 0;

  //定义并初始化灰度统计数组 hist[256]
  int hist[256];
  memset(hist, 0, sizeof(hist));

  //定义并初始化各灰度的概率数组 Freq[256]
  double Freq[256];
  // memset(Freq, 0. 0, sizeof(Freq));

  //定义类 1 概率总数, 类 2 概率总数
  double freq1, freq2;

  //定义类 1 熵, 类 2 熵, 总熵, KSW 熵
  double entropy1, entropy2, entropy, shannon;

  //最大熵值 max
  double max = 0. 0;

  //灰度最大、最小值
  int graymin = 255, graymax = 0;

```cpp
DWORD i, j;
int k;
if( wBitCount == 8) {
    //统计出各个灰度值的像素数
    for( i=0; i<nHeight; i++)
        for( j=0; j<nWidth; j++) {
            int gray = *(lpData+lRowBytes * i+j);
            hist[gray]++;
            //      Count++;
            if( gray>graymax)         graymax = gray;
            if( gray<graymin)         graymin = gray;
        }
    for( k=0; k<256; k++)    Freq[k] = 0.0;

    entropy = 0.0;
    entropy1 = 0.0;
    entropy2 = 0.0;
    shannon = 0.0;
    freq1 = 0.0;
    freq2 = 0.0;

    //求像素总数
    DWORD Count = nHeight * nWidth;

    //统计各灰度值的概率和总熵值
    for( k=graymin; k<graymax; k++) {
        if( hist[k]! = 0) {
            Freq[k] = hist[k] / (double)Count;
            entropy -= Freq[k] * log(Freq[k]) / log(E);
        }
    }
    //求阈值
    for( k=graymin; k<=graymax; k++) {
        //类1的概率
        freq1 += Freq[k];
        //类2概率
        freq2 = 1-freq1;
```

```cpp
                    if(Freq[k]!=0){
                        //类1熵
                        entropy1-=Freq[k]*log(Freq[k])/log(E);
                        //类2熵
                        entropy2=entropy-entropy1;
                        //KSW 熵
                        shannon=log(freq1*freq2)/log(E)+entropy1/freq1+entropy2/freq2;
                    }

                    if(shannon>max){
                        max=shannon;
                        nThreshold=k;
                    }
                }
            CString strInfo;
            strInfo.Format("阈值=%d\n", nThreshold);
            MessageBox(NULL, strInfo,"阈值结果", MB_OK);
        }
        ///////////////////////////////////////////
        //二值化图像
        for(i=0; i<nHeight; i++)
            for(j=0; j<nWidth; j++){
                if(*(lpData+lRowBytes*i+j)<nThreshold)   *(lpData+lRowBytes*i+j)=0;
                else             *(lpData+lRowBytes*i+j)=255;
            }
            return TRUE;
}
```

19. Hough 变换提取直线

/**
Hough 变换提取图像的直线

参数：
  Length   规定提取直线的长度
  *head   直线的起始链表指针
  *lineS   创建的节点指针
  *lineEnd   直线的尾端指针

返回：

BOOL　　　　　　　　　成功返回 TRUE，否则返回 FALSE。
***/
//存储 hough 变换得到的线段信息，如线段起始和结束端点

```cpp
struct MYLINE{
            int num;
            int topx;
            int topy;
            int botx;
            int boty;
            int angle;
            int dist;
            MYLINE *next;
            };
```

//记录对 k，l 的像素坐标，以及该像素对应的号码

```cpp
typedef struct{
            int x1, y1;
            int num;
            int p, q;
            int x2, y2;
            }KLPOINT;
```

BOOL HoughTransEdgeLink(int Length, MYLINE * head, MYLINE * lineS, MYLINE * lineEnd)
{
　　#define PI 3.1415926535
　　int　　　　　i, j, m, l, k;
　　int　　　　　　Dist, Alpha;
　　HGLOBAL　　　　hDistAlpha, hklpoint; //, hMyLine,;
　　int　　　　*lpDistAlpha;
　　KLPOINT　　　*lpklpoint, *temppoint, *temp;
　　int step = 1;
　　int l_max = 0;
　　int num_max = 0;
　　int DIST = 30; //此参数是用来判断两点是否距离很远，这样可能不属于一条线段

　　if(bitinfor = = NULL)　　　return FALSE;

```
DWORD nWidth = GetWidth();
DWORD nHeight = GetHeight();
WORD wBitCount = bitinfor->bmiHeader.biBitCount;
DWORD lRowBytes = WIDTHBYTES(nWidth * ((DWORD)wBitCount));
```

//运算区域
```
CRect r(0, 0, nWidth - 1, nHeight - 1);

Dist = (int)(sqrt((double)(r.bottom-r.top)*(r.bottom-r.top)+(double)(r.right-r.left)*(r.right-r.left)));
```

//步长
```
Alpha = 1800/step;  //0 degree to 180 degree, step is 1 degrees
```

//为存放数据分配空间,分配不成功返回 FALSE
```
if((hDistAlpha=GlobalAlloc(GHND, (DWORD)Dist * Alpha * sizeof(int)))==NULL){
    MessageBox(NULL,"Error alloc memory1!","Error Message", MB_OK | MB_ICONEXCLAMATION);
    return FALSE;
}

if((hklpoint=GlobalAlloc(GHND, (DWORD)Dist * Alpha * sizeof(KLPOINT)))==NULL){
    MessageBox(NULL,"Error alloc memory2!","Error Message", MB_OK | MB_ICONEXCLAMATION);
    GlobalFree(hDistAlpha);
    return FALSE;
}

lpDistAlpha = (int *)GlobalLock(hDistAlpha);
temp = lpklpoint = (KLPOINT *)GlobalLock(hklpoint);
```

//读取像素
```
BOOL bSuccess = TRUE;
temppoint = lpklpoint;
```

//扫描整个源图像
```
for (i = 0; i <= nHeight - 1; i++){
```

```cpp
        for(j = 0; j <= nWidth - 1; j++){
            LPBYTE pData = m_pDIBs + (nHeight - 1 - i) * lRowBytes + j;

            unsigned char c = *pData;

            if(c==0)   //如果灰度值为0,则认为是线段上一点
            {
                for(k=0; k<1800; k+=step)    //通过循环计算该点所属半径和角度
                {
                    //在各个方向上加上这个点出现的概率(次数)
                    //计算距离 l=x*cos(k)+y*sin(k)
            l=(int)fabs((j*cos(k*PI/1800.0)+(nHeight-i)*sin(k*PI/1800.0))); //PI=3.1415926535

                    if(l>l_max) l_max=l;

                    //通过距离为l,角度为k的像素个数(累加)
                    *(lpDistAlpha+l*Alpha+k/step) = *(lpDistAlpha+l*Alpha+k/step)+1;

                    if(*(lpDistAlpha+l*Alpha+k/step)>num_max)
                        num_max = *(lpDistAlpha+l*Alpha+k/step);

                    //将每一点的图像坐标以及半径、角度和序号(个数)存入KLPOINT结构体存储区域
                    if(*(lpDistAlpha+l*Alpha+k/step)<=2){//起点坐标

                        //如果是第一个点,则直接保存
                        if(*(lpDistAlpha+l*Alpha+k/step)==1){
                            (temppoint+l*Alpha+k/step)->x1=j;
                            (temppoint+l*Alpha+k/step)->y1=i;
                            (temppoint+l*Alpha+k/step)->num = *(lpDistAlpha+l*Alpha+k/step);
                            (temppoint+l*Alpha+k/step)->p=k;
                            (temppoint+l*Alpha+k/step)->q=l;

                        }
                        //如果是第二个点,则判断距离
                        if(*(lpDistAlpha+l*Alpha+k/step)==2){
```

```cpp
                    double distance=sqrt(pow((temppoint+l*Alpha+k/step)->x1-j,
2)+pow((temppoint+l*Alpha+k/step)->y1-i,2));

                    if(distance<DIST){
                        (temppoint+l*Alpha+k/step)->x2=j;
                        (temppoint+l*Alpha+k/step)->y2=i;
    (temppoint+l*Alpha+k/step)->num=*(lpDistAlpha+l*Alpha+k/step);
                        (temppoint+l*Alpha+k/step)->p=k;
                        (temppoint+l*Alpha+k/step)->q=l;
                    }
                    else{
                        (temppoint+l*Alpha+k/step)->x1=j;
                        (temppoint+l*Alpha+k/step)->y1=i;
    (temppoint+l*Alpha+k/step)->num=*(lpDistAlpha+l*Alpha+k/step)-1;
                        (temppoint+l*Alpha+k/step)->p=k;
                        (temppoint+l*Alpha+k/step)->q=l;
    *(lpDistAlpha+l*Alpha+k/step)=*(lpDistAlpha+l*Alpha+k/step)-1;
                    }}}
                else if(*(lpDistAlpha+l*Alpha+k/step)>2){//终点坐标
                    double distance=sqrt(pow((temppoint+l*Alpha+k/step)->x2-j,2)
+pow((temppoint+l*Alpha+k/step)->y2-i,2));
                    if(distance<DIST){
    (temppoint+l*Alpha+k/step)->num=*(lpDistAlpha+l*Alpha+k/step);
                        (temppoint+l*Alpha+k/step)->p=k;
                        (temppoint+l*Alpha+k/step)->q=l;
                        (temppoint+l*Alpha+k/step)->x2=j;
                        (temppoint+l*Alpha+k/step)->y2=i;
                    }
                    else{
                        (temppoint+l*Alpha+k/step)->num=*(lpDistAlpha+l*Alpha
+k/step)-1;
                        *(lpDistAlpha+l*Alpha+k/step)=*(lpDistAlpha+l*Alpha+
k/step)-1;
                    }}
                }//for(k)
            }//if
        }}
//根据提取线段长度将各线段存入链表
```

```cpp
temp=temppoint;
int w=0;
do{
    for( m=0; m<Dist; m=m+1)
        for (int n=0; n<1800; n+=10){
            if(((temppoint+m*Alpha+n/step)->num>Length \
                && (temppoint+m*Alpha+n/step)->num==num_max)
            {
                if(head==NULL)      head =lineS;
                else    lineEnd->next=lineS;  //结构体成员中有一个next成员
                lineEnd=lineS;
                lineS=new MYLINE;
                //起点坐标
                int x_start=(temppoint+m*Alpha+n/step)->x1;
                int y_start=(temppoint+m*Alpha+n/step)->y1;
                lineS->topx=x_start;
                lineS->topy=y_start;
                //终点坐标
                int x_end=(temppoint+m*Alpha+n/step)->x2;
                int y_end=(temppoint+m*Alpha+n/step)->y2;
                lineS->botx=x_end;
                lineS->boty=y_end;
                lineS->angle=n;
                lineS->dist=m;
                lineS->num=w++;
            }   //if
        }
    num_max--;
}while(num_max>Length);

lineEnd->next=NULL;
delete lineS;

GlobalUnlock(hDistAlpha);
GlobalFree(hDistAlpha);

GlobalUnlock(hklpoint);
GlobalFree(hklpoint);
```

```
    return bSuccess;
}
```

20. Huffman 编码

```
/***************************************************************
Huffman 编码
参数:
        FLOAT * * fFreq[in, out]      各个灰度值频率的数组指针
        WORD * iColorNum[in, out]     图像颜色数目
返回:        BOOL    成功编码返回 TRUE; 否则返回 FALSE
****************************************************************/
BOOL CodeHuffman(FLOAT * * fFreq, WORD * iColorNum)
{
    if(bitinfor = = NULL)    return FALSE;

    //图像像素总数
    LONG lCountSum;

    //临时计数器变量
    LONG i;
    LONG j;

    DWORD nWidth = GetWidth();
    DWORD nHeight = GetHeight();
    WORD wBitCount = bitinfor->bmiHeader.biBitCount;
    DWORD lRowBytes = WIDTHBYTES(nWidth * ((DWORD)wBitCount));
    LPBYTE lpData = m_pDIBs;
    LPBYTE lpOldBits;

    //获取当前图像的颜色数目
    * iColorNum = NumColors();

    //只处理 256 色位图的 Huffman 编码处理
    if( * iColorNum ! = 256){
        AfxMessageBox("目前只支持 256 色位图哈夫曼编码!");
        return FALSE;
    }
```

```cpp
    *fFreq = new FLOAT[*iColorNum];

    //初始化灰度值频率数组
    for(i = 0; i < *iColorNum; i ++)       (*fFreq)[i] = 0.0;

    //计算各个灰度值的计算(对于非256色位图,此处给数组 *fFreq 赋值方法将不同)
    for(i = 0; i < (WORD)nHeight; i ++){
        for(j = 0; j < (WORD)nWidth; j ++){
            lpOldBits = (unsigned char *)lpData + lRowBytes * i + j;

            // 计数加1
            (*fFreq)[*(lpOldBits)] += 1;
        }}

    //计算图像像素总数
    lCountSum = nHeight * nWidth;

    //计算各个灰度值出现的概率
    for(i = 0; i < *iColorNum; i ++)(*fFreq)[i] /=(FLOAT)lCountSum; // 计算概率
    return TRUE;
}
template <class T>
BOOL CodeHuffman(T, FLOAT * * fFreq, WORD * iColorNum)
{
    if (! IsValid())
    {
        return FALSE;
    }
    T *lpData = (T *)pData;
    double dMax, dMin;
    T *pp;

    //图像像素总数
    LONG lCountSum;

    //临时计数器变量
    LONG i;
    LONG j;
```

```cpp
    *fFreq = (FLOAT *)malloc(*iColorNum * sizeof(FLOAT));
    //初始化灰度值频率数组
    for(i = 0; i < *iColorNum; i ++)     (*fFreq)[i] = 0.0;

    for (int k = 0; k<nBandsShow; k++)
    {
        pDataSet->GetRasterBand(k+1)->GetStatistics(FALSE, TRUE, &dMin, &dMax, NULL, NULL);
        for(i = 0; i < (WORD)nHeight; i ++)
        {
            for(j = 0; j < (WORD)nWidth; j ++)
            {
                pp   = lpData + nWidth * i + j + k * nWidth * nHeight;

                // 计数加 1
                (*fFreq)[(BYTE)(*pp * 225/(dMax-dMin))] += 1;
            }
        }

    }

    //计算图像像素总数
    lCountSum = nHeight * nWidth;

    //计算各个灰度值出现的概率
    for(i = 0; i < *iColorNum; i ++)(*fFreq)[i] /= (FLOAT)lCountSum;  // 计算概率
    return TRUE;
}
```

21. 行程编码

```
/************************************************************
行程编码——图像对象保存为 256 色 PCX 文件
参数:
         CFile& file[out]     要保存的 PCX 文件的文件名
返回:    BOOL        函数是否调用成功
************************************************************/
BOOL DIBToPCX256(CFile &file)
{
```

```cpp
// PCX 文件头结构
typedef struct
{
    BYTE bManufacturer;
    BYTE bVersion;
    BYTE bEncoding;
    BYTE bBpp;
    WORD wLeft;
    WORD wTop;
    WORD wRight;
    WORD wBottom;
    WORD wXResolution;
    WORD wYResolution;
    BYTE bPalette[48];
    BYTE bReserved;
    BYTE bPlanes;
    WORD wLineBytes;
    WORD wPaletteType;
    WORD wSrcWidth;
    WORD wSrcDepth;
    BYTE bFiller[54];
}PCXHEADER;

//临时计数器循环变量
LONG i;
LONG j;

//定义高度和宽度
DWORD wHeight;
DWORD wWidth;

//相邻的两个像素(中间变量)
BYTE bChar1;
BYTE bChar2;

//指向原图像像素的指针
BYTE * lpSrc;
//指向编码后图像数据的指针
```

```
BYTE * lpDst;

//图像每行的字节数
LONG lLineBytes;

//重复像素计数
int iCount;

//缓冲区已经使用的字节数
DWORD dwBuffUsed;

//声明指向图像像素指针,并找到图像像素起始位置
LPBYTE lpDIBBits = m_pDIBs;

//获取图像高度
wHeight = GetHeight( );
//获取图像宽度
wWidth = GetWidth( );

//计算图像每行的字节数
lLineBytes = WIDTHBYTES( wWidth * 8 );

/* 给文件头赋值 */
// PCX 文件头
PCXHEADER pcxHdr;

// PCX 标识码
pcxHdr.bManufacturer = 0x0A;

// PCX 版本号
pcxHdr.bVersion = 5;

// PCX 编码方式(1 表示 RLE 编码)
pcxHdr.bEncoding = 1;

//像素位数(256 色为 8 位)
pcxHdr.bBpp = 8;
```

//图像相对于屏幕的左上角 X 坐标(以像素为单位)
pcxHdr. wLeft = 0;

//图像相对于屏幕的左上角 Y 坐标(以像素为单位)
pcxHdr. wTop = 0;

//图像相对于屏幕的右下角 X 坐标(以像素为单位)
pcxHdr. wRight = WORD(wWidth - 1);

//图像相对于屏幕的右下角 Y 坐标(以像素为单位)
pcxHdr. wBottom = WORD(wHeight - 1);

//图像的水平分辨率
pcxHdr. wXResolution = WORD(wWidth);

//图像的垂直分辨率
pcxHdr. wYResolution = WORD(wHeight);

//调色板数据(对于 256 色 PCX 无意义,直接赋值为 0)
for(i = 0; i < 48; i ++)   pcxHdr. bPalette[i] = 0;

//保留域,设定为 0
pcxHdr. bReserved = 0;

//图像色彩平面数目(对于 256 色 PCX 设定为 1)
pcxHdr. bPlanes = 1;

//图像的宽度(字节为单位),必须为偶数
if((wWidth & 1) = = 0)   pcxHdr. wLineBytes = WORD(wWidth);
else   pcxHdr. wLineBytes = WORD(wWidth + 1);

//图像调色板的类型,1 表示彩色或者单色图像,2 表示图像是灰度图
pcxHdr. wPaletteType = 1;

//制作该图像的屏幕宽度(像素为单位)
pcxHdr. wSrcWidth = 0;

//制作该图像的屏幕高度(像素为单位)

```
pcxHdr. wSrcDepth = 0;

//保留域,取值设定为0
for( i = 0; i < 54; i ++)      pcxHdr. bFiller[i] = 0;
```
//写入文件头
```
file. Write((LPSTR)&pcxHdr, sizeof( PCXHEADER));
```

/* 开始编码 */

//开辟一片缓冲区(2倍原始图像大小)以保存编码结果
```
lpDst = new BYTE[wHeight * wWidth * 2];
```

//指明当前已经用了多少缓冲区(字节数)
```
dwBuffUsed = 0;
```

//每行
```
for( i = 0; i < (LONG)wHeight; i ++){
    // 指向图像第i行,第0个像素的指针
    lpSrc = (BYTE *)lpDIBBits + lLineBytes * (wHeight -1 -i);

    // 给 bChar1 赋值
    bChar1 = *lpSrc;

    iCount = 1;

    // 剩余列
    for( j = 1; j < (LONG)wWidth; j ++){
        // 指向图像第i行,第j个像素的指针
        lpSrc++;

        // 读取下一个像素
        bChar2 = *lpSrc;

        // 判断是否和 bChar1 相同并且 iCount<63
        if((bChar1 == bChar2) && (iCount < 63)){
            // 相同,计数加1
            iCount++;
```

```cpp
                // 继续读下一个
            }
            else{
                // 不同, 或者 iCount = 63; 写入缓冲区
                if(((iCount > 1) || (bChar1 >= 0xC0))){
                    // 保存码长信息
                    lpDst[dwBuffUsed] = iCount | 0xC0;

                    // 保存 bChar1
                    lpDst[dwBuffUsed+1] = bChar1;

                    // 更新 dwBuffUsed
                    dwBuffUsed += 2;
                }
                else{
                    // 直接保存该值
                    lpDst[dwBuffUsed] = bChar1;

                    // 更新 dwBuffUsed
                    dwBuffUsed++;
                }

                // 重新给 bChar1 赋值
                bChar1 = bChar2;
                iCount = 1;
            }}

    // 保存每一行最后一部分编码
    if(((iCount > 1) || (bChar1 >= 0xC0))){
        // 保存码长信息
        lpDst[dwBuffUsed] = iCount | 0xC0;

        // 保存 bChar1
        lpDst[dwBuffUsed+1] = bChar1;

        // 更新 dwBuffUsed
        dwBuffUsed += 2;
    }
```

```cpp
        else{
            // 直接保存该值
            lpDst[dwBuffUsed] = bChar1;

            // 更新 dwBuffUsed
            dwBuffUsed++;
        }
    }

    //写入编码结果
    file.WriteHuge((LPSTR)lpDst, dwBuffUsed);

    //释放内存
    delete [] lpDst;

    /* 写入调色板信息 */
    //指向 BITMAPINFO 结构的指针
    LPBITMAPINFO lpbmi;

    //开辟一片缓冲区以保存调色板
    lpDst = new BYTE[769];

    //调色板起始字节
    *lpDst = 0x0C;

    //获取指向 BITMAPINFO 结构的指针
    lpbmi = bitinfor;

    //读取当前图像调色板
    for(i = 0; i < 256; i ++){
        // 读取调色板红色分量
        lpDst[i*3+1] = lpbmi->bmiColors[i].rgbRed;
        // 读取调色板绿色分量
        lpDst[i*3+2] = lpbmi->bmiColors[i].rgbGreen;
        // 读取调色板蓝色分量
        lpDst[i*3+3] = lpbmi->bmiColors[i].rgbBlue;
    }
    //写入调色板信息
    file.Write((LPSTR)lpDst, 769);
```

```
        delete [ ] lpDst;
        return TRUE;
}
```

22. 纹理图像的自相关函数分析法

```
/***************************************************************
纹理图像的自相关函数分析法
参数：      无
返回：      BOOL
***************************************************************/
BOOL Correlation( )
{
    //判断图像数据类型
    switch（datatype）
    {
    case GDT_Byte：
        Correlation(byte(0));
        break;
    case GDT_UInt16：
        Correlation(WORD(0));
        break;
    case GDT_Int16：
        Correlation(short(0));
        break;
    case GDT_UInt32：
        Correlation(DWORD(0));
        break;
    case GDT_Int32：
        Correlation(int(0));
        break;
    case GDT_Float32：
        Correlation(float(0));
        break;
    case GDT_Float64：
        Correlation(double(0));
        break;
    default：
        break;
    }
```

```cpp
    return TRUE;
}
template <class T>
BOOL Correlation(T)
{
    //判断图像是否为空
    if (! IsValid())
    {
        return FALSE;
    }
    //获取图像数据指针
    T *lpData = (T *)pData;
    //获取显示数据指针
    BYTE *pByte = m_pDIBs;
    //获取图像的宽度与高度
    DWORD nWidth = GetWidth();
    DWORD nHeight = GetHeight();
    //图像变换开始
    DWORD i, j, k, m, n;
    long double sum=0;
    long double *p = new long double [nHeight * nWidth * nBands];
    //初始化
    for (k = 0; k < nBands; k++)
    {
        for(m=0; m<nHeight; m++)
        {
            for(n=0; n<nWidth; n++)
            {
                *(p+nWidth*(m)+(n)+k*nWidth*nHeight)=0;
            }
        }
    }
    //逐波段进行处理
    for (k=0; k<nBands; k++)
    {
        sum = 0;
        for(i=0; i<nHeight; i++)
        {
```

```
                for(j=0; j<nWidth; j++)
                {
                    sum+= *(lpData +nWidth*(i)+(j)+k*nWidth*nHeight)*
( *(lpData +nWidth*(i)+(j))+k*nWidth*nHeight);
                }
            }
            //计算分子
            for(m=0; m<nHeight; m++)
            {
                for(n=0; n<nWidth; n++)
                {
                    for(i=0; i<nHeight; i++)
                    {
                        for(j=0; j<nWidth; j++)
                        {
                            if((i+m)<nHeight&&(j+n)<nWidth)
                                *(p+nWidth*(m)+(n)+k*nWidth*nHeight)+= *
(lpData +nWidth*(i)+(j)+k*nWidth*nHeight)*
                                    ( *(lpData +nWidth*(i+m)+(j+n)+k*nWidth*
nHeight));
                        }
                    }
                }
            }
            //分子除以分母
            for(i=0; i<nHeight; i++)
            {
                for(j=0; j<nWidth; j++)
                {
 *(p+nWidth*(i)+(j)+k*nWidth*nHeight)= *(p+nWidth*(i)+(j)+k*nWidth*
nHeight)/sum;
                }
            }
        }
        //修改显示数据，更新视图
        for (k=0; k<nBandsShow; k++)
        {
            //统计原图像最大和最小灰度级
```

```cpp
        float m_MinGray=(float)999999999, m_MaxGray=(float)-99999999;

        for (i=0; i<nHeight; i++)
        {
            for (j=0; j<nWidth; j++)
            {
                if(*(p+nWidth*(i)+(j)+k*nWidth*nHeight)<m_MinGray)
                    m_MinGray=*(lpData+nWidth*i+j+k*nWidth*nHeight);
                if(*(p+nWidth*(i)+(j)+k*nWidth*nHeight)>m_MaxGray)
                    m_MaxGray=*(lpData+nWidth*i+j+k*nWidth*nHeight);
            }
        }
        for (i=0; i<nHeight; i++)
        {
            for (j=0; j<nWidth; j++)
            {
                if (nBandsShow==1)
                    *(pByte+lLineBYTES*(nHeight-1-i)+j)=(BYTE)((*(p+nWidth*(i)+(j)+k*nWidth*nHeight)-m_MinGray)*255/(m_MaxGray-m_MinGray));
                else
                    *(pByte+lLineBYTES*(nHeight-1-i)+j*3+2-k)=(BYTE)((*(p+nWidth*(i)+(j)+k*nWidth*nHeight)-m_MinGray)*255/(m_MaxGray-m_MinGray));
            }
        }
    }
    delete []p;
    return TRUE;
}
```

23. 纹理图像的灰度共生矩阵分析法

```
/************************************************************
灰度共生矩阵法
参数：    void
返回：    BOOL
************************************************************/
BOOL GrayComatrix()
{
    //判断图像数据类型
    switch (datatype)
```

```
        }
        case GDT_Byte:
            GrayComatrix(byte(0));
            break;
        case GDT_UInt16:
            GrayComatrix(WORD(0));
            break;
        case GDT_Int16:
            GrayComatrix(short(0));
            break;
        case GDT_UInt32:
            GrayComatrix(DWORD(0));
            break;
        case GDT_Int32:
            GrayComatrix(int(0));
            break;
        case GDT_Float32:
            GrayComatrix(float(0));
            break;
        case GDT_Float64:
            GrayComatrix(double(0));
            break;
        default:
            break;
        }
        return TRUE;
}
template <class T>
BOOL GrayComatrix(T)
{
    //判断图像是否为空
    if(! IsValid())
    {
        return FALSE;
    }
    //获取图像数据指针
    T * lpData = (T *)pData;
    //获取显示数据指针
```

```cpp
BYTE *pByte = m_pDIBs;
//获取图像的宽度与高度
DWORD nWidth = GetWidth();
DWORD nHeight = GetHeight();
//分配内存
T *lpDataCopy=(T*)malloc(nHeight*nWidth*nBands);
if(lpDataCopy==NULL)
{
    //若内存申请失败,弹出消息并返回退出
    SetCursor(LoadCursor(NULL, IDC_ARROW));
    AfxMessageBox("Memory Allocate error");
    return FALSE;
}
//若内存分配成功,将位图数据拷贝到新申请的内存中
memcpy(lpDataCopy, lpData, nHeight*nWidth*nBandsShow*sizeof(T));
//灰度共生矩阵法获取纹理特征量

//共生矩阵
double p0[16][16];              //0度方向
double p45[16][16];             //45度方向
double p90[16][16];             //90度方向
double p135[16][16];            //135度方向

double f1[4];                   //角二阶矩
double f2[4];                   //惯性矩
double f3[4];                   //相关
double f4[4];                   //熵
double f5[4];                   //逆差矩

double u1[4], u2[4], delta1[4], delta2[4];   //计算相关时用到的量

double meanf1, meanf2, meanf3, meanf4, meanf5;   //5个特征量在4个方向的均值
meanf1=meanf2=meanf3=meanf4=meanf5=0;            //初始化

//正规化常数
long R0, R45, R90, R135;

DWORD i, j, k;
```

```cpp
    int m, n;

    //逐波段进行处理
    for (k=0; k<nBands; k++)
    {
        //初始化共生矩阵
        for(m=0; m<16; m++)
        {
            for(n=0; n<16; n++)
            {
                p0[m][n] = 0;
                p45[m][n] = 0;
                p90[m][n] = 0;
                p135[m][n] = 0;
            }
        }


        //初始化纹理特征量
        for(n=0; n<4; n++)
        {
            f1[n] = 0;
            f2[n] = 0;
            f3[n] = 0;
            f4[n] = 0;
            f5[n] = 0;

            u1[n] = 0;
            u2[n] = 0;
            delta1[n] = 0;
            delta2[n] = 0;
        }

        //计算图像的最大灰度级
        int pixel=0;
        double nMaxPixel = 0;
        for(i=0; i<nHeight; i++)
        {
            for(j=0; j<nWidth; j++)
```

```cpp
            }
            //找到最大灰度值
            if( *(lpData+nWidth*i+j+k*nWidth*nHeight)> nMaxPixel)
                nMaxPixel = *(lpData+nWidth*i+j+k*nWidth*nHeight);
        }
    }

//最大灰度级比最大灰度值至少要大1(如256级的最大值为255)
nMaxPixel = nMaxPixel + 1;

//求共生矩阵之前将灰度级压缩到16级
for(i=0; i<nHeight; i++)
{
    for(j=0; j<nWidth; j++)
    {

        //像素原来的灰度值
        pixel = *(lpData+nWidth*i+j+k*nWidth*nHeight);

        /* 灰度级压缩为16级
           注：nMaxPixel 必须大于 pixel，而不是大于或等于 pixel
           因为，如果某像素的灰度值 pixel = nMaxPixel ，则压缩后为灰度值
           16 此时会超出共生矩阵数组的最大下标15 */
        pixel = (int)( pixel * 16 / nMaxPixel);

        //给每个像素重新赋灰度值
        *(lpData+nWidth*i+j+k*nWidth*nHeight) = pixel;

    }
}

//当前像素的4个相邻方向的像素
int pixel_0, pixel_45, pixel_90, pixel_135;
//指向原图像的指针
T   *lpSrc;

//对4个方向分别求共生矩阵
for(i=0; i<nHeight; i++)
{
```

```
for(j=0; j<nWidth; j++)
{
    lpSrc=lpData+nWidth*i+j+k*nWidth*nHeight;

    //像素灰度值
    pixel = *lpSrc;

    // 求灰度共生矩阵

    //0度方向//防止越界
    if(j<nWidth-1)
    {
        //0度方向的相邻像素值(第i行第j+1个像素)
        pixel_0 = *(lpSrc+1);

        p0[pixel][pixel_0]++;
        p0[pixel_0][pixel]++;

    }

    //45度方向
    if(i<nHeight-1 && j<nWidth-1)
    {
        //间隔为1的45度方向的像素值(第i+1行第j+1个像素)
        pixel_45 = (BYTE)(*(lpSrc+nWidth+1));

        p45[pixel][pixel_45]++;
        p45[pixel_45][pixel]++;
    }

    //90度方向
    if(i<nHeight-1){
        //间隔为1的90度方向的像素值(第i+1行第j个像素)
        pixel_90 = (BYTE)(*(lpSrc+nWidth));

        //统计频率(共生矩阵)
        p90[pixel][pixel_90]++;
        p90[pixel_90][pixel]++;
```

```
            }

            //135 度方向
            if(j>0 && i<nHeight-1){
                //间隔为 1 的 135 度方向的像素值(第 i+1 行第 j-1 个像素)
                pixel_135 = (BYTE)( *(lpSrc+nWidth-1));

                //统计频率(共生矩阵)
                p135[pixel][pixel_135]++;
                p135[pixel_135][pixel]++;
            }

            //本来上面 2 个下标对称，只需求出一个，但当 pixel=nextpixel 时，需
要算作 2 次//故 2 种都统计
        }
    }

    //正规化处理
    R0 = 2 * nHeight * (nWidth-1);              //0 度方向正规化常数
    R45 = 2 * (nHeight-1) * (nWidth-1);         //45 度方向正规化常数
    R90 = 2 * (nHeight-1) * nWidth;             //90 度方向正规化常数
    R135 = 2 * (nHeight-1) * (nWidth-1);        //135 度方向正规化常数

    for(m=0; m<16; m++)
    {
        for(n=0; n<16; n++)
        {
            p0[m][n] = p0[m][n] / R0;           //正规化
            p45[m][n] = p45[m][n] / R45;        //正规化
            p90[m][n] = p90[m][n] / R90;        //正规化
            p135[m][n] = p135[m][n] / R135;     //正规化

            f1[0] += p0[m][n] * p0[m][n];       //求角二阶矩
            f1[1] += p45[m][n] * p45[m][n];     //求角二阶矩
            f1[2] += p90[m][n] * p90[m][n];     //求角二阶矩
            f1[3] += p135[m][n] * p135[m][n];   //求角二阶矩

            //计算 4 个方向的惯性矩 f2
```

```cpp
            f2[0] += (m-n)*(m-n)*p0[m][n];
            f2[1] += (m-n)*(m-n)*p45[m][n];
            f2[2] += (m-n)*(m-n)*p90[m][n];
            f2[3] += (m-n)*(m-n)*p135[m][n];
        }
    }

    //计算 u1
    double temp1, temp2, temp3, temp4;              //中间变量
    temp1 = temp2 = temp3 = temp4 = 0;

    for(m=0; m<16; m++)
    {
        temp1 = temp2 = temp3 = temp4 = 0;
        for(n=0; n<16; n++)
        {
            temp1 += p0[m][n];
            temp2 += p45[m][n];
            temp3 += p90[m][n];
            temp4 += p135[m][n];
        }
        u1[0] += temp1 * m;
        u1[1] += temp2 * m;
        u1[2] += temp3 * m;
        u1[3] += temp4 * m;
    }

    //计算 u2
    for(n=0; n<16; n++)
    {
        temp1 = temp2 = temp3 = temp4 = 0;
        for(m=0; m<16; m++)
        {
            temp1 += p0[m][n];
            temp2 += p45[m][n];
            temp3 += p90[m][n];
            temp4 += p135[m][n];
        }
```

```cpp
            u2[0] += temp1 * n;
            u2[1] += temp2 * n;
            u2[2] += temp3 * n;
            u2[3] += temp4 * n;
        }
    //计算delta1
    for(m=0; m<16; m++)
    {
        temp1 = temp2 = temp3 = temp4 = 0;
        for(n=0; n<16; n++)
        {
            temp1 += p0[m][n];
            temp2 += p45[m][n];
            temp3 += p90[m][n];
            temp4 += p135[m][n];
        }
        delta1[0] += temp1 * (m-u1[0]) * (m-u1[0]);
        delta1[1] += temp2 * (m-u1[1]) * (m-u1[1]);
        delta1[2] += temp3 * (m-u1[2]) * (m-u1[2]);
        delta1[3] += temp4 * (m-u1[3]) * (m-u1[3]);
    }

    //计算delta2
    for(n=0; n<16; n++)
    {
        temp1 = temp2 = temp3 = temp4 = 0;
        for(m=0; m<16; m++)
        {
            temp1 += p0[m][n];
            temp2 += p45[m][n];
            temp3 += p90[m][n];
            temp4 += p135[m][n];
        }
        delta2[0] += temp1 * (n-u2[0]) * (n-u2[0]);
        delta2[1] += temp2 * (n-u2[1]) * (n-u2[1]);
        delta2[2] += temp3 * (n-u2[2]) * (n-u2[2]);
        delta2[3] += temp4 * (n-u2[3]) * (n-u2[3]);
    }
```

```cpp
//计算4个方向的相关f3
temp1 = temp2 = temp3 = temp4 = 0;
for(m=0; m<16; m++)
{
    for(n=0; n<16; n++)
    {
        temp1 += m * n * p0[m][n];
        temp2 += m * n * p45[m][n];
        temp3 += m * n * p90[m][n];
        temp4 += m * n * p135[m][n];
    }
}


f3[0] = (temp1-u1[0]*u2[0]) / (delta1[0]*delta2[0]);
f3[1] = (temp2-u1[1]*u2[1]) / (delta1[1]*delta2[1]);
f3[2] = (temp3-u1[2]*u2[2]) / (delta1[2]*delta2[2]);
f3[3] = (temp4-u1[3]*u2[3]) / (delta1[3]*delta2[3]);

//计算4个方向的熵f4和逆差矩f5
for(m=0; m<16; m++)
{
    for(n=0; n<16; n++)
    {
        //熵f4
        if(p0[m][n] > 0)     //取对数时真数部分必须大于0
            f4[0] -= p0[m][n] * log(p0[m][n])/log(2.0);

        if(p45[m][n] > 0)
            f4[1] -= p45[m][n] * log(p45[m][n])/log(2.0);

        if(p90[m][n] > 0)
            f4[2] -= p90[m][n] * log(p90[m][n])/log(2.0);

        if(p135[m][n] > 0)
            f4[3] -= p135[m][n] * log(p135[m][n])/log(2.0);

        //逆差矩f5
        f5[0] += p0[m][n] /(1+(m-n)*(m-n));
```

```
                    f5[1] += p45[m][n] /(1+(m-n)*(m-n));
                    f5[2] += p90[m][n] /(1+(m-n)*(m-n));
                    f5[3] += p135[m][n] /(1+(m-n)*(m-n));
                }
            }

            //计算4个方向的平均值
            for(m=0; m<4; m++)
            {
                meanf1 += f1[m];
                meanf2 += f2[m];
                meanf3 += f3[m];
                meanf4 += f4[m];
                meanf5 += f5[m];
            }
            meanf1 = meanf1 / 4.0;
            meanf2 = meanf2 / 4.0;
            meanf3 = meanf3 / 4.0;
            meanf4 = meanf4 / 4.0;
            meanf5 = meanf5 / 4.0;

            //消息框显示结果
            CString strInfo;
            strInfo.Format("第%d 波段 \n 二阶矩=%.3lf \n 惯性矩=%.3lf \n 相关=%.3lf \n 熵=%.3lf \n 逆差矩=%.3lf", k+1, meanf1, meanf2, meanf3, meanf4, meanf5);
            MessageBox(NULL, strInfo,"灰度共生矩阵特征量", MB_OK);

        }
        //由于求共生矩阵时对图像进行了灰度级压缩, 此处将图像还原
        memcpy(lpData, lpDataCopy, nBands*nHeight*nWidth);

        return TRUE;
    }
```

24. 二值图像形状特征提取
该程序适用于处理8位二值图像。
/***
二值图像形状特征提取

参数：无
返回：BOOL
**/
```cpp
BOOL ShapeFeatures()
{
    typedef struct
    {
        int index;
        CPoint lefttop;
        CPoint rightbottom;
        int area;
        int arealength;
        float rectangularfit;
        int barea;
        int roundfit;
        int shapeindex;
        int barealength;
        int lengthtowide;
    } ShapeFeature;
    int num;
    ShapeFeature * m_shapefeature = NULL;
    if(bitinfor == NULL)    return FALSE;

    //获得实际的位图数据占用的字节数
    DWORD m_dwSizeImage = bitinfor->bmiHeader.biSizeImage;

    LPBYTE m_lpImageCopy = (BYTE *)malloc(m_dwSizeImage);
    if(m_lpImageCopy == NULL){
        SetCursor(LoadCursor(NULL, IDC_ARROW));
        AfxMessageBox("Memory Allocate error");
        return FALSE;
    }

    memcpy(m_lpImageCopy, m_pDIBs, m_dwSizeImage);

    DWORD nWidth = GetWidth();
    DWORD nHeight = GetHeight();
    WORD wBitCount = bitinfor->bmiHeader.biBitCount;
```

```cpp
DWORD lRowBytes = WIDTHBYTES(nWidth * ((DWORD)wBitCount));
LPBYTE lpData = m_pDIBs;
LPBYTE lpOldBits = m_lpImageCopy;

int stop=0;
int counter=0;
int present;
DWORD i, j, m, n, t;
BYTE * p_temp;
p_temp=new BYTE[lRowBytes * nHeight];
memset(p_temp, 255, lRowBytes * nHeight);

const int T=50;
if(wBitCount==8){
for(i=0; i<nWidth; i++)        *(lpData+(nHeight-1) * lRowBytes+i)= 255;
for(j=0; j<nHeight; j++)       *(lpData+(nHeight-1-j) * lRowBytes)= 255;

for(j=1; j<nHeight-1; j++){
    if(stop==1)
        break;
    for(i=1; i<nWidth-1; i++){
        if(counter>255){
            AfxMessageBox("连通区域数目太多,请减少样本个数");
            stop=1;
            return FALSE;
        }
        if(*(lpData+(nHeight-j-1) * lRowBytes+i)<T) {
            if(*(lpData+(nHeight-j-1+1) * lRowBytes+i+1)<T) {
    *(p_temp+(nHeight-j-1) * lRowBytes+i)= *(p_temp+(nHeight-j-1+1) * lRowBytes+i+1);
                present = *(p_temp+(nHeight-j-1+1) * lRowBytes+i+1);
    if(*(lpData+(nHeight-j-1) * lRowBytes+i-1)<T&& *(p_temp+(nHeight-j-1) * lRowBytes+i-1)! =present)
{
                    int temp= *(p_temp+(nHeight-j-1) * lRowBytes+i-1);
                    if(present>temp) {
                        present=temp;
                        temp= *(p_temp+(nHeight-j-1+1) * lRowBytes+i+1);
```

```
                    }
                    counter--;
                    for(m=1; m<nHeight; m++)
                        for(n=1; n<nWidth; n++){
                            if(*(p_temp+(nHeight-m-1)*lRowBytes+n)==temp)
                                {
                                    *(p_temp+(nHeight-m-1)*lRowBytes+n)=present;
                                }
                            else if(*(p_temp+(nHeight-m-1)*lRowBytes+n)>temp){
                                    *(p_temp+(nHeight-m-1)*lRowBytes+n)-=1;
                            }}}//end 左前
    if(*(lpData+(nHeight-j-1+1)*lRowBytes+i-1)<T&&*(p_temp+(nHeight-j-1+1)
*lRowBytes+i-1)!=present){
                    counter--;
                    int temp=*(p_temp+(nHeight-j-1+1)*lRowBytes+i-1);
                    if(present<temp)  {
                        present=temp;
                        temp=*(p_temp+(nHeight-j-1+1)*lRowBytes+i-1);
                    }

                    for(m=1; m<nHeight; m++)
                        for(n=1; n<nWidth; n++){
                            if(*(p_temp+(nHeight-m-1)*lRowBytes+n)==
present){
                                    *(p_temp+(nHeight-m-1)*lRowBytes+n)=temp;
                                }
                            else if(*(p_temp+(nHeight-m-1)*lRowBytes+n)>
present){
                                    *(p_temp+(nHeight-m-1)*lRowBytes+n)-=1;
                                }}
                    present=temp;
            }//end 左上
        }
                else if(*(lpData+(nHeight-j-1+1)*lRowBytes+i)<T)  {
    *(p_temp+(nHeight-j-1)*lRowBytes+i)=*(p_temp+(nHeight-j-1+1)*lRowBytes
+i);
                    present=*(p_temp+(nHeight-j-1+1)*lRowBytes+i);
```

```
                    else if( *(lpData+(nHeight-j-1+1)*lRowBytes+i-1)<T) {
*(p_temp+(nHeight-j-1)*lRowBytes+i)= *(p_temp+(nHeight-j-1+1)*lRowBytes+i-1);
                    present= *(p_temp+(nHeight-j-1+1)*lRowBytes+i-1);
                }
                    else if( *(lpData+(nHeight-j-1)*lRowBytes+i-1)<T)  {
*(p_temp+(nHeight-j-1)*lRowBytes+i)= *(p_temp+(nHeight-j-1)*lRowBytes+i-1);
                    present= *(p_temp+(nHeight-j-1)*lRowBytes+i-1);
                }
                    else{
                        ++counter;
                        present=counter;
                        *(p_temp+(nHeight-1-j)*lRowBytes+i)=present;
                }}}}}
    num=counter;

    if(m_shapefeature! =NULL)
        delete []m_shapefeature;

    m_shapefeature=new ShapeFeature[num];
    for(i=0; i<num; i++)  {
        m_shapefeature[i].index=i+1;
        m_shapefeature[i].lefttop.x=nWidth;
        m_shapefeature[i].lefttop.y=nHeight;
        m_shapefeature[i].rightbottom.x=0;
        m_shapefeature[i].rightbottom.y=0;
        m_shapefeature[i].area=0;
        m_shapefeature[i].arealength=0;
        m_shapefeature[i].rectangularfit=0.0;
        m_shapefeature[i].barea=0;
        m_shapefeature[i].roundfit=0;
        m_shapefeature[i].shapeindex=0;
    }
    for(t=1; t<num+1; t++)  {
        for(j=1; j<nHeight-1; j++)  {
            for(i=1; i<nWidth-1; i++)  {
                if( *(p_temp+(nHeight-j-1)*lRowBytes+i)= =t)  {
                    if(m_shapefeature[t-1].lefttop.x>i)
```

```
                    m_shapefeature[t-1].lefttop.x=i;
                if(m_shapefeature[t-1].lefttop.y>j)
                    m_shapefeature[t-1].lefttop.y=j;
                if(m_shapefeature[t-1].rightbottom.x<i)
                    m_shapefeature[t-1].rightbottom.x=i;
                if(m_shapefeature[t-1].rightbottom.y<j)
                    m_shapefeature[t-1].rightbottom.y=j;
            }}}}
//////////////////////计算面积
        for(t=0; t<num; t++)
            for(j=1; j<nHeight-1; j++)
                for(i=1; i<nWidth-1; i++)
                    if(*(p_temp+(nHeight-j-1)*lRowBytes+i)==t+1)    {
                        m_shapefeature[t].area++;
                    }
//////////////////////计算周长
        for(j=1; j<nHeight-1; j++)
            for(i=1; i<nWidth-1; i++)
                if(*(lpData+j*lRowBytes+i)-*(lpData+j*lRowBytes+i+1)==255)
                    *(lpOldBits+j*lRowBytes+i+1)=100;
                else if (*(lpData+j*lRowBytes+i+1)-*(lpData+j*lRowBytes+i)==255)
                    *(lpOldBits+j*lRowBytes+i)=100;
                else if(*(lpData+j*lRowBytes+i)-*(lpData+(j+1)*lRowBytes+i)==255)
                    *(lpOldBits+(j+1)*lRowBytes+i)=100;
                else if(*(lpData+(j+1)*lRowBytes+i)-*(lpData+j*lRowBytes+i)==255)
                    *(lpOldBits+j*lRowBytes+i)=100;

        for( t=0; t<num; t++)
            for(i=1; i<nWidth-1; i++)
                for(j=1; j<nHeight-1; j++)
                    if(*(p_temp+j*lRowBytes+i)==t+1)    {
                        if(*(lpOldBits+j*lRowBytes+i)==100)
                            m_shapefeature[t].arealength++;
                    }
//////////////////计算其他形状特征
        for(t=0; t<num; t++){
m_shapefeature[t].barea=abs((m_shapefeature[t].lefttop.x-m_shapefeature[t].rightbottom.x-1)
```

```
                        *(m_shapefeature[t].lefttop.y-m_shapefeature[t].
rightbottom.y-1));
m_shapefeature[t].barealength=2*abs((m_shapefeature[t].lefttop.x-m_shapefeature[t].
rightbottom.x-1)+2*abs(m_shapefeature[t].lefttop.y-m_shapefeature[t].rightbottom.y
-1));

m_shapefeature[t].shapeindex=(m_shapefeature[t].arealength*m_shapefeature[t].
arealength)/(m_shapefeature[t].area*4*3.14159);

m_shapefeature[t].rectangularfit=(double)m_shapefeature[t].area/m_shapefeature[t].
barea;

m_shapefeature[t].lengthtowide=(double)abs(m_shapefeature[t].lefttop.x-m_shapefeature
[t].rightbottom.x-1)/abs(m_shapefeature[t].lefttop.y-m_shapefeature[t].rightbottom.
y-1);

m_shapefeature[t].roundfit=(double)(m_shapefeature[t].arealength*m_shapefeature[t].
arealength)/m_shapefeature[t].area;}

    CString *str1 = new CString[num];
    CString *str2 = new CString[num];
    CString *str3 = new CString[num];

    CString s1;
    for(i=0;i<num;i++) {
    s1.Format("代号%d 面积%d \n",
    m_shapefeature[i].index,
    m_shapefeature[i].area);
    str1[i]+=s1;
    }
    CString s2;
    for(i=0;i<num;i++) {
    s2.Format("代号%d 周长%d \n",
    m_shapefeature[i].index,
    m_shapefeature[i].arealength);
    str2[i]+=s2;
    }
    CString s3;
```

```
        for(i=0; i<num; i++) {
s3.Format("代号%d 圆形度%f \ n",
m_shapefeature[i].index,
m_shapefeature[i].roundfit);
str3[i]+=s3;
       }
       return TRUE;
}
```

25. 基于灰度的模板匹配

```
/ *********************************************************
基于灰度的模板匹配
参数:
    double     *lpTemplateData       指向模板图像指针
    DWORD nTemplateWidth             模板图像宽度(像素数)
    DWORD nTemplateHeight            模板图像高度(像素数)
返回:
    BOOL       运算成功返回 TRUE,否则返回 FALSE。
**********************************************************/
BOOLTemplateMatchDIB ( double * lpTemplateData, DWORD nTemplateWidth, DWORD nTemplateHeight)
{
    //判断图像数据类型
    switch (datatype)
    {
    case GDT_Byte:
        TemplateMatchDIB(byte(0), lpTemplateData, nTemplateWidth, nTemplateHeight);
        break;
    case GDT_UInt16:
        TemplateMatchDIB(WORD(0), lpTemplateData, nTemplateWidth, nTemplateHeight);
        break;
    case GDT_Int16:
        TemplateMatchDIB(short(0), lpTemplateData, nTemplateWidth, nTemplateHeight);
        break;
    case GDT_UInt32:
        TemplateMatchDIB(DWORD(0), lpTemplateData, nTemplateWidth, nTemplateHeight);
        break;
    case GDT_Int32:
        TemplateMatchDIB(int(0), lpTemplateData, nTemplateWidth, nTemplateHeight);
```

```cpp
            break;
        case GDT_Float32:
            TemplateMatchDIB(float(0), lpTemplateData, nTemplateWidth, nTemplateHeight);
            break;
        case GDT_Float64:
            TemplateMatchDIB(double(0), lpTemplateData, nTemplateWidth, nTemplateHeight);
            break;
        default:
            break;
    }
    returnTRUE;
}
template <class T>
BOOL TemplateMatchDIB(T, double * lpTemplateData, DWORD nTemplateWidth, DWORD nTemplateHeight)
{
    //判断图像是否为空
    if(! IsValid())
    {
        return FALSE;
    }
    //获取图像数据指针
    T * lpData = (T *)pData;

    //获取图像的宽度与高度
    DWORD nWidth = GetWidth();
    DWORD nHeight = GetHeight();
    //指向原始图像的指针
    T * lpSrc;
    double * lpTemplateSrc;

    T * lpDst;
    //循环变量
    DWORD i;
    DWORD j;
    DWORD m;
    DWORD n;
```

```cpp
    //中间结果
    double dSigmaST;
    double dSigmaS;
    double dSigmaT;

    //相似性测度
    double R;

    //最大相似性测度
    double MaxR;

    //最大相似性出现位置
    long lMaxWidth;
    long lMaxHeight;

    //像素值
    double pixel;
    double templatepixel;

    //计算 dSigmaT
    dSigmaT = 0;
    for (n = 0; n < nTemplateHeight; n++)
    {
        for(m = 0; m < nTemplateWidth; m++)
        {
            // 指向模板图像倒数第 j 行,第 i 个像素的指针
            lpTemplateSrc = lpTemplateData + nTemplateWidth * n + m;
            templatepixel = * lpTemplateSrc;
            dSigmaT += (double)templatepixel * templatepixel;
        }
    }

//找到图像中最大相似性的出现位置
MaxR = 0.0;
for (j = 0; j < nHeight-nTemplateHeight +1; j++)
{
    for(i = 0; i < nWidth-nTemplateWidth + 1; i++)
    {
```

```
                dSigmaST = 0;
                dSigmaS = 0;

                for ( n = 0; n < nTemplateHeight ; n++){
                    for( m = 0; m < nTemplateWidth ; m++){
                        // 指向源图像倒数第 j+n 行, 第 i+m 个像素的指针
                        lpSrc    = lpData + nWidth * (j+n) + (i+m);

                        // 指向模板图像倒数第 n 行, 第 m 个像素的指针
                        lpTemplateSrc    = lpTemplateData + nTemplateWidth * n + m;

                        pixel = ( * lpSrc);
                        templatepixel = * lpTemplateSrc;

                        dSigmaS += (double)pixel * pixel;
                        dSigmaST += (double)pixel * templatepixel;
                    }}
                //计算相似性
                R = dSigmaST / ( sqrt(dSigmaS) * sqrt(dSigmaT));
                //与最大相似性比较
                if ( R > MaxR )
                {
                    MaxR = R;
                    lMaxWidth = i;
                    lMaxHeight = j;
                }
            }
        }
        //将最大相似性出现区域部分复制到目标图像
        for ( n = 0; n < nTemplateHeight ; n++) {
            for( m = 0; m < nTemplateWidth ; m++) {
                lpTemplateSrc = lpTemplateData + nTemplateWidth * n + m;
                lpDst = lpData + nWidth * (n+lMaxHeight) + (m+lMaxWidth);
                * lpDst = (T) * lpTemplateSrc;
            }}

        //消息框显示结果
        CString strInfo;
```

```
    strInfo.Format("匹配点坐标:(%d,%d)", lMaxWidth, lMaxHeight);
    MessageBox(NULL, strInfo,"影像匹配", MB_OK);
    return TRUE;
}
```

26. 基于特征的模板匹配

```
/***********************************************************
基于特征的模板匹配
参数:
        double * lpTemplateData         指向模板图像指针
        DWORD nTemplateWidth            模板图像宽度(像素数)
        DWORD nTemplateHeight           模板图像高度(像素数)
返回:   BOOL                            运算成功返回 TRUE,否则返回 FALSE。
***********************************************************/
BOOL TemplateMatch(double * lpTemplateData, DWORD nTemplateWidth, DWORD nTemplateHeight)
{
    //判断图像数据类型
    switch(datatype)
    {
    case GDT_Byte:
        TemplateMatch(byte(0), lpTemplateData, nTemplateWidth, nTemplateHeight);
        break;
    case GDT_UInt16:
        TemplateMatch(WORD(0), lpTemplateData, nTemplateWidth, nTemplateHeight);
        break;
    case GDT_Int16:
        TemplateMatch(short(0), lpTemplateData, nTemplateWidth, nTemplateHeight);
        break;
    case GDT_UInt32:
        TemplateMatch(DWORD(0), lpTemplateData, nTemplateWidth, nTemplateHeight);
        break;
    case GDT_Int32:
        TemplateMatch(int(0), lpTemplateData, nTemplateWidth, nTemplateHeight);
        break;
    case GDT_Float32:
        TemplateMatch(float(0), lpTemplateData, nTemplateWidth, nTemplateHeight);
        break;
    case GDT_Float64:
```

```cpp
            TemplateMatch(double(0), lpTemplateData, nTemplateWidth, nTemplateHeight);
            break;
        default:
            break;
        }
        return TRUE;
}
template <class T>
BOOL TemplateMatch(T, double * lpTemplateData, DWORD nTemplateWidth, DWORD nTemplateHeight)
{
    //判断图像是否为空
    if(! IsValid())
    {
        return FALSE;
    }
    //获取图像数据指针
    T * lpData = (T *)pData;
    //获取图像的宽度与高度
    DWORD nWidth = GetWidth();
    DWORD nHeight = GetHeight();
    //指向原始图像的指针
    double * lpSrc;
    double * lpTemplateSrc;
    double * lpDataSrc = new double[nWidth * nHeight];

    //循环变量
    DWORD i;
    DWORD j;
    DWORD m;
    DWORD n;

    for(j = 0; j < nHeight; j++)
    {
        for(i = 0; i < nWidth; i++)
        {
            lpDataSrc[j * nWidth+i] = (double)( *(lpData+j * nWidth+i));
        }
```

}
//调用 Sobel( )函数对图像进行边缘检测
Sobel(lpDataSrc, nWidth, nHeight);
Sobel(lpTemplateData, nTemplateWidth, nTemplateHeight);

//像素值
double pixel;
double templatepixel;

DWORD min = INT_MAX;
DWORD T = 10; //阈值

CRect matchedArea;
//pDataSet->GetRasterBand(1)->GetStatistics(FALSE, TRUE, &dMin, &dMax, NULL, NULL);
for (j = 0; j < nHeight-nTemplateHeight +1; ++j)
{
    for(i = 0; i < nWidth-nTemplateWidth +1; ++i)
    {
        DWORD d = 0;
        int flag = -1;

        //和模板求最小距离
        for(n = 1; n < nTemplateHeigh -1; ++n)//模板行
        {
            for(m = 1; m <nTemplateWidth -1 ; ++m)//模板列
            {
                // 指向源图像倒数第 j+n 行, 第 i+m 个像素的指针
                lpSrc   = lpDataSrc + nWidth * (j+n) + (i+m);

                // 指向模板图像倒数第 n 行, 第 m 个像素的指针
                lpTemplateSrc   = lpTemplateData + nTemplateWidth * n+m;

                pixel = (*lpSrc);
                templatepixel = *lpTemplateSrc;
                d += abs(pixel-templatepixel);
            }//end of m

```
                    if(d>T)
                    {
                        flag=1;
                        break;
                    }
                }//end of n

                if(flag == 1)   continue;
                if(min > d)
                {
                    min = d;
                    if(min<T)
                    {
                        //矩形框显示区域
                        matchedArea.left = i;
                        matchedArea.top = j;
                        matchedArea.right = i+nTemplateWidth-1;
                        matchedArea.bottom = j+nTemplateHeight-1;
                    }
                }
        }//end of i
    }//end of j
    //消息框显示结果
    CString strInfo;
    strInfo.Format("匹配区域：\n 上:%d\n 下:%d\n 左:%d\n 右:%d",
matchedArea.top, matchedArea.bottom, matchedArea.left, matchedArea.right);
    MessageBox(NULL, strInfo,"影像匹配", MB_OK);
    return TRUE;
}
```

/**

Sobel 算子检测边缘

参数：

  double * lpData   指向图像指针

  DWORD wWidth   图像宽度(像素数)

  DWORD wHeight   图像高度(像素数)

返回： BOOL   运算成功返回 TRUE，否则返回 FALSE。

***/

BOOL Sobel(double * lpData, WORD wWidth, WORD wHeight)

```cpp
}
    double *m_lpImageCopy = new double[wWidth * wHeight];
    if(m_lpImageCopy == NULL)
    {
        //若内存申请失败,弹出消息并返回退出
        SetCursor(LoadCursor(NULL, IDC_ARROW));
        AfxMessageBox("Memory Allocate error");
        return FALSE;
    }
    //图像变换开始
    DWORD i, j, m, n;
    double sum;
    int t, Sobel[9];

    //若内存分配成功,将位图数据拷贝到新申请的内存中
    memcpy(m_lpImageCopy, lpData, wWidth * wHeight * sizeof(double));

    //设置 Sobel 算子
    Sobel[0] = -1;
    Sobel[1] =  0;
    Sobel[2] =  1;
    Sobel[3] = -2;
    Sobel[4] =  0;
    Sobel[5] =  2;
    Sobel[6] = -1;
    Sobel[7] =  0;
    Sobel[8] =  1;

    for(i=1; i<wHeight-1; i++)
    {
        for(j=1; j<wWidth-1; j++)
        {
            sum=0; t=0;
            //3×3 模板算子运算
            for(m=0; m<3; m++)
            {
                for(n=0; n<3; n++)
                {
```

```
                        sum+= * (m_lpImageCopy +wWidth * (i-1+m)+(j-1+n)) * Sobel[t++];
                    }
                    if(sum<0)                * (lpData+wWidth * (i)+(j)) = 0;
                    else if(sum>255)         * (lpData+wWidth * (i)+(j)) = 255;
                    else                     * (lpData+wWidth * (i)+(j)) = sum;
                }
            }
        }
    return TRUE;
}
```

27. 基于高通滤波的图像融合

本程序适用于处理已配准的同一地区多光谱图像和高分辨率全色图像的融合。

```
/***************************************************************
基于高通滤波的图像融合
参数：   float * Pan_Data            全色图像数据
返回：void
****************************************************************/
void HPFNONWEIGHT(float * Pan_Data)
{
    //判断多光谱图像数据类型
    switch (datatype)
    {
    case GDT_Byte:
        HPFNONWEIGHT(byte(0), Pan_Data);
        break;
    case GDT_UInt16:
        HPFNONWEIGHT(WORD(0), Pan_Data);
        break;
    case GDT_Int16:
        HPFNONWEIGHT(short(0), Pan_Data);
        break;
    case GDT_UInt32:
        HPFNONWEIGHT(DWORD(0), Pan_Data);
        break;
    case GDT_Int32:
        HPFNONWEIGHT(int(0), Pan_Data);
        break;
```

```cpp
        case GDT_Float32:
            HPFNONWEIGHT(float(0), Pan_Data);
            break;
        case GDT_Float64:
            HPFNONWEIGHT(double(0), Pan_Data);
            break;
        default:
            break;
    }
}
template <class T>
void HPFNONWEIGHT(T, float * Pan_Data)
{
    //判断图像是否为空
    if (! IsValid())
    {
        return;
    }
    int i, j, k;
    //获取图像数据指针
    T * lpData = (T *)pData;
    //获取显示数据指针
    BYTE * pByte = m_pDIBs;
    //获取图像的宽度与高度
    DWORD nWidth = GetWidth();
    DWORD nHeight = GetHeight();
    float * MSData = new float[nWidth * nHeight * nBands];
    //拷贝源数据
    for (k = 0; k < nBands; k++)
    {
        for (j = 0; j < nHeight; j++)
        {
            for (i = 0; i < nWidth; i++)
            {
                MSData[j*nWidth+i+k * nWidth * nHeight] = lpData[j * nWidth+i+k * nWidth * nHeight];
            }
        }
```

```cpp
    }
    int Division = 1;
    int Offset = 0;
    double values[9][9];
    int Size = 3;
    values[0][0] = values[0][2] = values[2][0] = values[2][2] = 0;
    values[0][1] = values[1][0] = values[2][1] = values[1][2] = -1;
    values[1][1] = 4;
    //开辟与图像同样大小的数组,用于保存图像结果
    float *pPanTemp = new float[nWidth*nHeight];
    for(i = 0 ; i < nHeight; i++)
    {
        for(j = 0; j < nWidth; j++)
        {
            if (i == 0 || i == nHeight-1 || j == 0 || j == nWidth-1)
            {
                *(pPanTemp+i*nWidth+j) = *(Pan_Data+i*nWidth+j);
            }
            else
            {
                double gray = 0;
                double sum = 0.0;
                double cof=0; double size=0;
                double c = 0;
                for(int m=0; m<Size; m++)
                {
                    for(int n=0; n<Size; n++)
                    {
                        float *pTemp = Pan_Data+(i-Size/2+m)*nWidth+(j-Size/2+n);
                        sum += (values[m][n] * (*pTemp));
                        c   += values[m][n];
                        size ++;
                    }//end n
                }//end m
                gray = sum;
                cof = c;
                if(cof <0)    //如果系数小于0,取原值
```

```cpp
            {
                float *pTemp = Pan_Data + i*nWidth+j;
                gray = *pTemp;
            }
            else if(size <Size*Size)
            {
                if(cof = = 0) cof =1;
                gray = min(gray/cof +Offset, 255);
            }
            else gray = min(gray/Division + Offset, 255);
            gray = max(0, gray);
            *(pPanTemp+i*nWidth+j) = gray;
        }
    }
}
//逐波段进行处理
for (k = 0; k < nBands; k++)
{
    for( i = 0; i < nHeight; i++ ){
        for(j = 0; j < nWidth; j++)  {
            float *t1    = pPanTemp + i*nWidth+j;
            float *t2    = MSData + i*nWidth+j+k*nWidth*nHeight;
            if (datatype = = GDT_Byte)
            {
                *(lpData+i*nWidth+j+k*nWidth*nHeight) = max(0, min(*t1
+ *t2, 255));
            }
            else
                *(lpData+i*nWidth+j+k*nWidth*nHeight) = (T)(*t1+*t2);
        }}
}
//修改显示数据,更新视图
for (k=0; k<nBandsShow; k++)
{
    //统计原图像最大和最小灰度级
    float m_MinGray=(float)999999999, m_MaxGray=(float)-99999999;
    if (datatype = = GDT_Byte)
    {
```

```
                    m_MaxGray = 255;
                    m_MinGray = 0;
                }
            else
                {
                    for (i=0; i<nHeight; i++)
                    {
                        for (j=0; j<nWidth; j++)
                        {
                            if( *(lpData+nWidth*i+j+k*nWidth*nHeight)<m_MinGray)
                                m_MinGray = *(lpData+nWidth*i+j+k*nWidth*nHeight);
                            if( *(lpData+nWidth*i+j+k*nWidth*nHeight)>m_MaxGray)
                                m_MaxGray = *(lpData+nWidth*i+j+k*nWidth*nHeight);
                        }
                    }
                }
            for (i=0; i<nHeight; i++)
            {
                for (j=0; j<nWidth; j++)
                {
                    if (nBandsShow == 1)
        *(pByte+lLineBYTES*(nHeight-1-i)+j) = (BYTE)((*(lpData+nWidth*i+j+k*
nWidth*nHeight)-m_MinGray)*255/(m_MaxGray-m_MinGray));
                    else
        *(pByte+lLineBYTES*(nHeight-1-i)+j*3+2-k) = (BYTE)((*(lpData+nWidth*i
+j+k*nWidth*nHeight)-m_MinGray)*255/(m_MaxGray-m_MinGray));
                }
            }
        }
    delete []MSData;   delete []pPanTemp;
}
```

28. 高频调制融合法

本程序适用于处理已配准的同一地区多光谱图像和高分辨率全色图像的融合。
/***
图像的高频调制融合法
参数：
  float *Pan_Data   全色图像数据
返回：   void

```cpp
    ******************************************************/
void HPFWEIGHT(float *Pan_Data)
{
    //判断多光谱图像数据类型
    switch (datatype)
    {
    case GDT_Byte:
        HPFWEIGHT(byte(0), Pan_Data);
        break;
    case GDT_UInt16:
        HPFWEIGHT(WORD(0), Pan_Data);
        break;
    case GDT_Int16:
        HPFWEIGHT(short(0), Pan_Data);
        break;
    case GDT_UInt32:
        HPFWEIGHT(DWORD(0), Pan_Data);
        break;
    case GDT_Int32:
        HPFWEIGHT(int(0), Pan_Data);
        break;
    case GDT_Float32:
        HPFWEIGHT(float(0), Pan_Data);
        break;
    case GDT_Float64:
        HPFWEIGHT(double(0), Pan_Data);
        break;
    default:
        break;
    }
}
template <class T>
void HPFWEIGHT(T, float *Pan_Data)
{
    //判断图像是否为空
    if (!IsValid())
    {
        return;
```

```cpp
    }
    int i, j, k;
    //获取图像数据指针
    T *lpData = (T *)pData;
    //获取显示数据指针
    BYTE *pByte = m_pDIBs;
    //获取图像的宽度与高度
    DWORD nWidth = GetWidth();
    DWORD nHeight = GetHeight();
    float *MSData = new float[nWidth*nHeight*nBands];
    //拷贝源数据
    for (k = 0; k < nBands; k++)
    {
        for (j = 0; j < nHeight; j++)
        {
            for (i = 0; i < nWidth; i++)
            {
                MSData[j*nWidth+i+k*nWidth*nHeight] = lpData[j*nWidth+i+k*nWidth*nHeight];
            }
        }
    }
    int Division = 1;
    int Offset = 0;
    double values[9][9];
    int Size = 3;
    values[0][0] = values[0][1] = values[0][2] = 1;
    values[1][0] = values[1][1] = values[1][2] = 1;
    values[2][0] = values[2][1] = values[2][2] = 1;
    for( i = 0; i < 3; i++)
        for(int j = 0; j < 3; j++)   values[i][j] /= 9;

    //开辟与图像同样大小的数组,用于保存图像结果
    float *pPanTemp = new float[nWidth*nHeight];
    for(i = 0 ; i < nHeight; i++)
    {
        for(j = 0; j < nWidth; j++)
        {
```

```cpp
if (i = = 0 || i = = nHeight-1 || j = = 0 || j = = nWidth-1)
{
    *(pPanTemp+i*nWidth+j) = *(Pan_Data+i*nWidth+j);
}
else
{
    double gray = 0;
    double sum = 0.0;
    double cof=0; double size=0;
    double c = 0;
    for(int m=0; m<Size; m++)
    {
        for(int n=0; n<Size; n++)
        {
            float *pTemp=Pan_Data+ (i-Size/2+m)*nWidth + (j-Size/2+n);
            sum += (values[m][n] * (*pTemp));
            c    += values[m][n];
            size ++;
        }//end n
    }//end m
    gray = sum;
    cof = c;
    if(cof <0)   //如果系数小于0，取原值
    {
        float *pTemp = Pan_Data + i*nWidth+j;
        gray = *pTemp;
    }
    else if(size <Size*Size)
    {
        if(cof = = 0) cof = 1;
        gray = min(gray/cof +Offset, 255);
    }
    else gray = min(gray/Division + Offset, 255);
    gray = max(0, gray);
    *(pPanTemp+i*nWidth+j) = gray;
}
}
}
```

```cpp
//逐波段进行处理
for (k = 0; k < nBands; k++)
{
    for( i = 0; i < nHeight; i++ ){
        for(j = 0; j < nWidth; j++)  {
            float * now    = pPanTemp + i * nWidth+j;
            float * row = Pan_Data + i * nWidth+j;
            float * pData2   = MSData + i * nWidth+j+k * nWidth * nHeight;
            float no = (float) * now;
            float ro =  * row;
            float high = ro-no;
            float fTemp =  * pData2+ * pData2 * (high/no);
            if (datatype == GDT_Byte)
            {
                * (lpData+i * nWidth+j+k * nWidth * nHeight) = max(0, min (fTemp, 255));
            }
            else
                * (lpData+i * nWidth+j+k * nWidth * nHeight) = (T)(fTemp);
        }}
}
//修改显示数据，更新视图
for (k=0; k<nBandsShow; k++)
{
    //统计原图像最大和最小灰度级
    float m_MinGray=(float)999999999, m_MaxGray=(float)-99999999;
    if (datatype == GDT_Byte)
    {
        m_MaxGray = 255;
        m_MinGray = 0;
    }
    else
    {
        for (i=0; i<nHeight; i++)
        {
            for (j=0; j<nWidth; j++)
            {
                if( * (lpData+nWidth * i+j+k * nWidth * nHeight)<m_MinGray)
```

```
                    m_MinGray = *(lpData+nWidth * i+j+k * nWidth * nHeight);
                if(*(lpData+nWidth * i+j+k * nWidth * nHeight)>m_MaxGray)
                    m_MaxGray = *(lpData+nWidth * i+j+k * nWidth * nHeight);
            }
        }
    }
    for (i=0; i<nHeight; i++)
    {
        for (j=0; j<nWidth; j++)
        {
            if (nBandsShow == 1)
*(pByte+lLineBYTES*(nHeight-1-i)+j) = (BYTE)((*(lpData+nWidth*i+j+k*nWidth*nHeight)-m_MinGray)*255/(m_MaxGray-m_MinGray));
            else
*(pByte+lLineBYTES*(nHeight-1-i)+j*3+2-k) = (BYTE)((*(lpData+nWidth*i+j+k*nWidth*nHeight)-m_MinGray)*255/(m_MaxGray-m_MinGray));
        }
    }
}
delete []MSData;    delete []pPanTemp;
}
```

29. HIS 变换融合法

本程序适用于处理已配准的同一地区多光谱图像和高分辨率全色图像的融合。

```
/*************************************************************
HIS 变换融合法
参数：      float * HighData         高分辨率图像数据
返回：      void
*************************************************************/
void IHSfusion(float * HighData)
{
    //判断多光谱图像数据类型
    switch (datatype)
    {
    case GDT_Byte:
        IHSfusion(byte(0), HighData);
        break;
    case GDT_UInt16:
        IHSfusion(WORD(0), HighData);
```

```cpp
            break;
        case GDT_Int16:
            IHSfusion(short(0), HighData);
            break;
        case GDT_UInt32:
            IHSfusion(DWORD(0), HighData);
            break;
        case GDT_Int32:
            IHSfusion(int(0), HighData);
            break;
        case GDT_Float32:
            IHSfusion(float(0), HighData);
            break;
        case GDT_Float64:
            IHSfusion(double(0), HighData);
            break;
        default:
            break;
    }
}
template <class T>
void IHSfusion(T, float * HighData)
{
    //判断图像是否为空
    if (! IsValid() || nBandsShow<3)
    {
        return;
    }
    int i, j, k;
    //获取图像数据指针
    T * lpData = (T *)pData;
    //获取显示数据指针
    BYTE * pByte = m_pDIBs;
    //获取图像的宽度与高度
    DWORD nWidth = GetWidth();
    DWORD nHeight = GetHeight();
    float * MSData = new float[nWidth * nHeight * nBands];
    for (k = 0; k < 3; k++)
```

```cpp
    }
        for (j = 0; j < nHeight; j++)
        {
            for (i = 0; i < nWidth; i++)
            {
                MSData[j*nWidth+i+k*nWidth*nHeight] = lpData[j*nWidth+i+k*nWidth*nHeight];
            }
        }
}
float * Intensity  = new float[nHeight*nWidth];
float * Hue        = new float[nHeight*nWidth];
float * Saturation = new float[nHeight*nWidth];
//对多光谱图像进行 ISH 变换
TransformfromRGBtoIHS(MSData, Intensity, Hue, Saturation);
//高分辨率图像与 I 进行匹配
Histogrammatch(Intensity, HighData);
//对多光谱图像进行 ISH 逆变换
TransformfromIHStoRGB(MSData, Intensity, Hue, Saturation);
for (k = 0; k < 3; k++)
{
    for( i = 0; i < nHeight; i++ ){
        for(j = 0; j < nWidth; j++)    {
            float fTemp = *(MSData+i*nWidth+j+k*nWidth*nHeight);
            if (datatype == GDT_Byte)
            {
                *(lpData+i*nWidth+j+k*nWidth*nHeight) = max(0, min(fTemp, 255));
            }
            else
                *(lpData+i*nWidth+j+k*nWidth*nHeight) = (T)(fTemp);
        }}
}
//修改显示数据,更新视图
for (k=0; k<nBandsShow; k++)
{
    //统计原图像最大和最小灰度级
    float m_MinGray=(float)999999999, m_MaxGray=(float)-99999999;
```

```
            if (datatype == GDT_Byte)
            {
                m_MaxGray = 255;
                m_MinGray = 0;
            }
            else
            {
                for (i=0; i<nHeight; i++)
                {
                    for (j=0; j<nWidth; j++)
                    {
                        if( *(lpData+nWidth*i+j+k*nWidth*nHeight)<m_MinGray)
                            m_MinGray = *(lpData+nWidth*i+j+k*nWidth*nHeight);
                        if( *(lpData+nWidth*i+j+k*nWidth*nHeight)>m_MaxGray)
                            m_MaxGray = *(lpData+nWidth*i+j+k*nWidth*nHeight);
                    }
                }
            }
            for (i=0; i<nHeight; i++)
            {
                for (j=0; j<nWidth; j++)
                {
                    if (nBandsShow == 1)
  *(pByte+lLineBYTES*(nHeight-1-i)+j) = (BYTE)((*(lpData+nWidth*i+j+k*
nWidth*nHeight)-m_MinGray)*255/(m_MaxGray-m_MinGray));
                    else
  *(pByte+lLineBYTES*(nHeight-1-i)+j*3+2-k) = (BYTE)((*(lpData+nWidth*i
+j+k*nWidth*nHeight)-m_MinGray)*255/(m_MaxGray-m_MinGray));
                }
            }
        }
        delete []MSData;
        delete []Intensity;
        delete []Hue;
        delete []Saturation;
        Intensity    = NULL;
        Hue          = NULL;
```

```
    Saturation = NULL;
}
/*************************************************************
对多光谱图像进行 ISH 变换
参数:      float  * rgb            RGB 色彩空间
           float  * Intensity      强度
           float  * Hue            色调
           float  * Saturation     饱合度
返回:      void
*************************************************************/
void TransformfromRGBtoIHS( float  * rgb, float  * Intensity, float  * Hue, float  * Saturation)
{
    float r, g, b, I, H, S;

    //获取图像的宽度与高度
    DWORD nWidth = GetWidth( );
    DWORD nHeight = GetHeight( );
    for( int i=0; i<nHeight; i++)
    {
        for( int j=0; j<nWidth; j++)
        {
            r = *( rgb   + nWidth * i + j + nWidth * nHeight * 2);
            g = *( rgb   + nWidth * i + j + nWidth * nHeight);
            b = *( rgb   + nWidth * i + j);
            double V1, V2;
            V1 = (2 * r-g-b)/sqrt(6);
            V2 = (g-b)/sqrt(2);
            I = float((r+g+b)/sqrt(3));
            float m = min(min(r, g), b);
            S = float( 1 - sqrt(3) * m/I);
            double f = 0.5 * (2 * r-g-b)/sqrt((r-g) * (r-g)+(r-b) * (g-b));
            if(g>=b)    H = float( acos(f));
            else        H = float( 2 * PI-acos(f));
            //保存
            Intensity[i * nWidth+j]  = I;
            Hue[i * nWidth+j]        = H;
            Saturation[i * nWidth+j] = S;
        } // end for j
```

} // end for i
}
/ ***
对多光谱图像进行 ISH 逆变换
参数： float * rgb   RGB 色彩空间
       float * Intensity   强度
       float * Hue   色调
       float * Saturation   饱合度
返回： void
**/
void TransformfromIHStoRGB(float * rgb, float * Intensity, float * Hue, float * Saturation)
{
    if((Intensity = = NULL))
    {
        AfxMessageBox("无正变换，错误!", 0, 0) ;
        return ;
    }
    float I, H, S;
    float R, G, B;

    //获取图像的宽度与高度
    DWORD nWidth = GetWidth( ) ;
    DWORD nHeight = GetHeight( ) ;
    for(int i = 0; i<nHeight; i++)
    {
        for(int j = 0; j<nWidth; j++)
        {
            I = Intensity[i * nWidth+j] ;
            H = Hue[i * nWidth+j] ;
            S = Saturation[i * nWidth+j] ;
            if(H>=0&&H<2 * PI/3)
            {
                R = I * (1+S * cos(H)/cos(PI/3-H))/sqrt(3) ;
                B = I * (1-S)/sqrt(3) ;
                G = sqrt(3) * I-R-B;
            }
            else if(H>=2 * PI/3&&H<4 * PI/3)
            {
```
```

```
                G = I*(1+S*cos(H-2*PI/3)/cos(PI-H))/sqrt(3);
                R = I*(1-S)/sqrt(3);
                B = sqrt(3)*I-R-G;
            }
            else if(H>=4*PI/3&&H<2*PI)
            {
                B = I*(1+S*cos(H-4*PI/3)/cos(5*PI/3-H))/sqrt(3);
                G = I*(1-S)/sqrt(3);
                R = sqrt(3)*I-B-G;
            }
            *(rgb   + nWidth*i + j) = B;
            *(rgb   + nWidth*i + j + nWidth*nHeight) = G;
            *(rgb   + nWidth*i + j + nWidth*nHeight*2) = R;
        } // end for j
    } // end for i
}
/*************************************************************
高分辨率图像与I进行匹配
参数：   float * pPrincipalImageI      第一主分量
         float * HighData              高分辨率图像数据
返回：   void
*************************************************************/
void Histogrammatch(float * pPrincipalImageI, float * HighData)
{
    //赋初值
    double MinHigh, MaxHigh, MinI, MaxI;
    MinHigh = MinI = 100000;
    MaxHigh = MaxI = 0;
    float Panch, Principal;
    //计算两幅图像的最大最小值
    for(int i=0; i<nHeight; i++){
        for(int j=0; j<nWidth; j++){
            //高分辨率图像
            float *pTemp = HighData+ i*nWidth+j;
            Panch = (float)*pTemp;
            if(Panch>MaxHigh) MaxHigh = Panch;
            if(Panch<MinHigh) MinHigh = Panch;
```

```
            //第一主分量
            Principal = pPrincipalImageI[i*nWidth+j];
            if(Principal>MaxI) MaxI = Principal;
            if(Principal<MinI) MinI = Principal;
        }//end for j
    }//end for i
    //拉伸高分辨率图像,结果保存在第一主分量中

    for(i=0; i<nHeight; i++)
    {
        for(int j=0; j<nWidth; j++)
        {
            float *pTemp = HighData+ i*nWidth+j;
            Panch = (double)*pTemp;
            Panch    = double((Panch-MinHigh)*(MaxI-MinI)/(MaxHigh-MinHigh)
+MinI);
            pPrincipalImageI[i*nWidth+j] = Panch;
        }
    }
}
```

30. 白平衡算法

```
void CImage::WhiteBalance(CImage *Img)
{
    int height=Img->GetHeight();
    int width=Img->GetWidth();
    int channels=3; //只处理 RGB 彩色图像
    int stepImg = width*3;
    BYTE* dataImg= (BYTE*)malloc(sizeof(BYTE)*width*height*3);
    BYTE* dataImgResult= (BYTE*)Img->m_pDIBs;
    memcpy(dataImg, dataImgResult, height*width*3);

    double *Y=new double[height*width];
    double YAve=0;
    int i, j;
    for (i=0; i<height; i++)
    {
        for (j=0; j<width; j++)
        {
```

```cpp
            Y[i*width+j] = 0.299 * dataImg[i*stepImg+j*channels+2] +
                0.587 * dataImg[i*stepImg+j*channels+1] +
                0.114 * dataImg[i*stepImg+j*channels+0];
            YAve+ = Y[i*width+j];
        }
    }
    YAve = YAve/(height*width);

    double RAve=0, GAve=0, BAve=0;
    int num=0;
    for (i=0; i<height; i++)
    {
        for (j=0; j<width; j++)
        {
            RAve+ = dataImg[i*stepImg+j*channels+2];
            GAve+ = dataImg[i*stepImg+j*channels+1];
            BAve+ = dataImg[i*stepImg+j*channels+0];
            num++;
        }
    }
    RAve = RAve/num;
    GAve = GAve/num;
    BAve = BAve/num;

    double K[3];
    K[2] = YAve/RAve;
    K[1] = YAve/GAve;
    K[0] = YAve/BAve;

    double maxK = 0;
    maxK = K[0]<K[1] ? K[1] : K[0];
    maxK = maxK<K[2] ? K[2] : maxK;

    K[0] = K[0]/maxK;
    K[1] = K[1]/maxK;
    K[2] = K[2]/maxK;

    for (i=0; i<height; i++)
```

```cpp
        {
            for (j=0; j<width; j++)
            {
                int temp=i*stepImg+j*channels;
                for (int m=0; m<3; m++)
                {
                    dataImgResult[temp+m]=(BYTE)(K[m]*dataImg[temp+m]);
                }
            }
        }
    delete [] Y;
}
```

31. 基于SUSAN算子的角点提取

该程序适用于8位灰度图像

/***
SUSAN算子角点提取
参数：无
返回： void
***/

```cpp
void Susan()
{
if(m_pBMI==NULL)
    //获得实际的位图数据占用的字节数
        DWORD m_dwSizeImage=m_pBMI->bmiHeader.biSizeImage;
    LPBYTE m_lpImageCopy=(BYTE *)malloc(m_dwSizeImage);
    if(m_lpImageCopy==NULL)
    {
        SetCursor(LoadCursor(NULL, IDC_ARROW));
        AfxMessageBox("Memory Allocate error");
        return FALSE;
    }
    memcpy(m_lpImageCopy, m_pBits, m_dwSizeImage);

    DWORD nWidth=Width();
    DWORD nHeight=Height();
        WORD wBitCount=m_pBMI->bmiHeader.biBitCount;
    DWORD lRowBytes=WIDTHBYTES(nWidth*((DWORD)wBitCount));
    DWORD step=lRowBytes/sizeof(BYTE);
```

```cpp
    LPBYTE data0=m_pBits;
    LPBYTE data1=m_lpImageCopy;

    DWORD i, j, k, same, max, min, thresh, sum;
    //x方向模板和y方向模板 37*37
    int OffSetX[37] = { -1, 0, 1,
        -2, -1, 0, 1, 2,
        -3, -2, -1, 0, 1, 2, 3,
        -3, -2, -1, 0, 1, 2, 3,
        -3, -2, -1, 0, 1, 2, 3,
        -2, -1, 0, 1, 2,
        -1, 0, 1 };
    int OffSetY[37] = { -3, -3, -3,
        -2, -2, -2, -2, -2,
        -1, -1, -1, -1, -1, -1, -1,
        0, 0, 0, 0, 0, 0, 0,
        1, 1, 1, 1, 1, 1, 1,
        2, 2, 2, 2, 2,
        3, 3, 3 };
//
    max = min = *(data0);
    for(i=3; i<nHeight-3; i++)
        for(j=3; j<nWidth-3; j++)
        {
            same =0; sum = 0;
            for(k=0; k<37; k++)
            {
                sum+= *(data1+(i+OffSetY[k])*step+j+OffSetX[k]);
                thresh = sum/37;
    if(fabs( *(data1+(i+OffSetY[k])*step+j+OffSetX[k])- *(data1+(i)*step+j))<=thresh)
                    same++;
                if(same<18)
                    *(data0+lRowBytes*(i)+(j))= 255;
                else
                    *(data0+lRowBytes*(i)+(j))= 0;
            }
        }
```

return TRUE;
}

32. K 均值聚类算法

参数：pData1，pData2，pData3//3 个波段的图像数据

Cols，rows 分别代表图像的列号和行号

class_number 代表分类的类别数，由键盘输入

```
//仅以 3 波段影像为例
//----------------------------定义类的成员函数--------------------//
//计算新的聚类中心
void center(vector<vector<Point>> y, double * z, int class_number)//double * z = new double (class_number * 3)
{
//分别计算各类群的均值向量
    for(int i=0; i<class_number; i++)
    {
        for(int j=0; i<y.at(i).size(); j++)
        {
            z[i*3+0]+=(y.at(i)).at(j).R*1.0/y.at(i).size();
            z[i*3+1]+=(y.at(i)).at(j).G*1.0/y.at(i).size();
            z[i*3+2]+=(y.at(i)).at(j).B*1.0/y.at(i).size();
        }
    }
}

int min(double * temp, int class_number)
{
    int flag = 0;
    int i, j, _temp;
    for(i=0; i<class_number; i++)
    {
        if(temp[flag]<temp[i])
            flag++;
    }
    return flag;
}

//以最小距离原则分类：计算各样本到各聚类中心的距离并按最小距离分类存放在各聚类数组中
void classify(BYTE x1[], BYTE x2[], BYTE x3[], double * z, vector<vector<Point>
```

```cpp
> y, int class_number, int *cal_class, int cols, int rows)
{
    Point *data = new Point[cols*rows];
    int i, j;
    for (i=0; i<rows; i++)
    {
        for (j=0; j<cols; j++)
        {
            data[i*cols+j].R = x1[i*cols+j];
            data[i*cols+j].G = x2[i*cols+j];
            data[i*cols+j].B = x3[i*cols+j];
        }
    }
    int count = cols*rows;
    double *temp = new double[count*class_number];
    double t;
    Int *count = new int(class_number);  //各个类别的样本总数
    for(i=0; i<count; i++)
    {
        //分别计算各样本到各聚类中心的距离
        for(j=0; j<class_number; j++)
        {
            t=(x1[i]-z[j][0])*(x1[i]-z[j][0])+(x2[i]-z[j][1])*(x2[i]-z[j][1])+(x3[i]-z[j][2])*(x3[i]-z[j][2]);
            temp[i*class_number+j]=sqrt(t);
        }
        //寻找样本到三个聚类中心的距离的最小值
        cal_class[i] = min(temp, class_number);
    }
}
//--------------------以下为主函数----------------------------//
//读取图像，此处省略代码
//确定初始聚类中心
double *OriCluster = new double(class_number*3) = {……此处学生自定义……}; //用来存储原始聚类中心
double *z = new double(class_number*3); //用来存储新的聚类中心
Vector<vector<Point>> y;
int *cal_class = new int[cols*rows];
```

319

```
double *temp = new temp(class_number*3);
double flag = 1;
while(flag)//判断最近两次迭代结果的聚类中心是否一致
{
        //调用函数进行聚类
        classify(pData1, pData2, pData3, OriCluster, y, class_number, cal_class,
    cols, rows);
        //调用函数计算新的聚类中心
            center(y, z, class_number);
            //计算新的聚类中心与上一次的聚类中心的差值
        for(i=0; i<class_number; i++)
            for(j=0; j<3; j++)
                temp[i*3+j]=z[i*3+j]-OriCluster[i*3+j];
        for(int i=0; i<class_number*3)
            flag*=temp[i];

        //将新的聚类中心存放于初始聚类中心数组中
for(i=0; i<class_number; i++)
    for(j=0; j<3; j++)
        {
OriCluster[i*3+j]=z[i*3+j];
z[i*3+j]=0;
}
}
```

33. 分水岭分割

```
/******************************************************************
利用 OpenCV 进行分水岭分割
参数：
IplImage *src 输入影像
IplImage *dst 输出影像
int Window_Size 窗口
Sections& sections 区域
返回值：bool  函数调用是否成功。若成功则返回 true，否则返回 false
******************************************************************/
const int WATERSHED = -1;
bool watershed(IplImage *src, IplImage *dst, int Window_Size, Sections& sections)
{
```

```cpp
    if ( ! src || ! dst)
    {
        return false;
    }
    IplImage * gray = NULL;
    if ( src->nChannels == 3)
    {
        gray = cvCreateImage( cvSize( src->width, src->height), IPL_DEPTH_8U, 1);
        cvCvtColor( src, gray, CV_RGB2GRAY);
    }
    else if ( src->nChannels == 1)
    {
        gray = cvCloneImage( src);
    }
    else
    {
        return false;
    }
    IplImage * median = cvCreateImage( cvSize( src->width, src->height), IPL_DEPTH_8U, 1);
    cvSmooth( gray, median, CV_MEDIAN, 3);
    cvReleaseImage( &gray);
    double * gradient = new double[ src->width * src->height];
    getGradient( median, gradient);
    cvReleaseImage( &median);
    vector<pair<int, int> > seedPoints;
    int * label = new int[ src->width * src->height];
    searchSeedPoints( gradient, src->width, src->height, seedPoints, Window_Size, label);
    double lambda[10] = {0.1, 0.2, 0.3, 0.4, 0.5, 0.6, 0.7, 0.8, 0.9, 1.0};
    double bata[10] = {2.5, 5, 7.5, 10, 12.5, 15, 17.5, 20, 22.5, 25};
    waterFill( lambda[3], bata[3], seedPoints, gradient, Window_Size, src->height, src->width, label);
    sections.findSections( label, src->height, src->width);
    setColor( dst, label);
    cvNamedWindow("分割结果", CV_WINDOW_AUTOSIZE);
    cvShowImage("分割结果", dst);
    cvWaitKey(0);
```

```
            cvDestroyWindow("分割结果");
            delete[] gradient;
            delete[] label;
            return true;
}
/ ****************************************************************
计算影像梯度
参数：
    IplImage  *src 输入影像
    double  *gradient 影像梯度
返回值：无
*****************************************************************/
        void getGradient(IplImage *src, double *gradient)
        {
            int i=0, j=0;
            double pixel[9]={0};
            double dx=0, dy=0;
            for (i = 1; i < src->height-1; i++)
            {
                for (j = 1; j < src->width-1; j++)
                {
                    pixel[0]=CV_IMAGE_ELEM(src, uchar, i-1, j-1);
                    pixel[1]=CV_IMAGE_ELEM(src, uchar, i-1, j);
                    pixel[2]=CV_IMAGE_ELEM(src, uchar, i-1, j+1);
                    pixel[3]=CV_IMAGE_ELEM(src, uchar, i, j-1);
                    pixel[4]=CV_IMAGE_ELEM(src, uchar, i, j);
                    pixel[5]=CV_IMAGE_ELEM(src, uchar, i, j+1);
                    pixel[6]=CV_IMAGE_ELEM(src, uchar, i+1, j-1);
                    pixel[7]=CV_IMAGE_ELEM(src, uchar, i+1, j);
                    pixel[8]=CV_IMAGE_ELEM(src, uchar, i+1, j+1);
                    dx=pixel[2]+2*pixel[5]+pixel[8]-pixel[0]-2*pixel[3]-pixel[6];
                    dx=dx/4;
                    dy=pixel[0]+2*pixel[1]+pixel[2]-pixel[6]-2*pixel[7]-pixel[8];
                    dy=dy/4;
                    gradient[i*src->width+j]=sqrt(dx*dx+dy*dy);
                }
            }
            int pos=0;
```

```cpp
    for (i=0; i<src->height; i++)
    {
        pos=i*src->width;
        gradient[pos]=gradient[pos+1];
        pos=i*src->width+src->width-1;
        gradient[pos]=gradient[pos-1];
    }
    for (j=0; j<src->width; j++)
    {
        pos=j;
        gradient[pos]=gradient[pos+src->width];
        pos=(src->height-1)*src->width+j;
        gradient[pos]=gradient[pos-src->width];
    }
}
/******************************************************************
寻找种子点
参数：
    double *gradient 输入影像梯度
    int width      影像宽
    int height     影像高
    vector<pair<int, int> > &seedPoints 种子点
    int Window_Size 窗口大小
    int *label 编号
返回值：无
******************************************************************/
void searchSeedPoints(double *gradient, int width, int height, vector<pair<int, int> > &seedPoints, int Window_Size , int *label)
{
    if (!gradient || !label)
    {
        return;
    }
    seedPoints.clear();
    pair<int, int> tempPos;
    int cols=width;
    int rows=height;
    int num=cols*rows;
```

```cpp
for (int i=0; i<num; ++i)
{
    label[i]=WATERSHED-1;
}
int searchWidth=5;
if(Window_Size<searchWidth)
{
    Window_Size=searchWidth;
}
double minGradient=500;
int seedNum=0;
for(int r=Window_Size/2; r<rows; r+=Window_Size/2)
{
    for(int c=Window_Size/2; c<cols; c+=Window_Size/2)
    {
        minGradient=500;
        int r_flag=0, c_flag=0;
        for(int i=-searchWidth/2; i<=searchWidth/2; i++)
        {
            for(int j=-searchWidth/2; j<=searchWidth/2; j++)
            {
                int rr=r+i;
                int cc=c+j;
                if(rr<rows&&cc<cols&& gradient[rr*cols+cc]<minGradient)
                {
                    minGradient=gradient[rr*cols+cc];
                    r_flag=rr;
                    c_flag=cc;
                }
            }
        }
        label[r_flag*cols+c_flag]=seedNum++;
        tempPos.first=r_flag;
        tempPos.second=c_flag;
        seedPoints.push_back(tempPos);
    }
}
```

/ **
涨水子函数
　　参数：
　　　　int center 中心
　　　　int pos 点的位置
　　　　int * label 编号
返回值：无
* **/
```
void expandBasin(int center, int pos, int * label)
{
    if ( label[pos]>WATERSHED )
    {
        if (label[center]<WATERSHED)
        {
            label[center] = label[pos];
        }
        else if (label[center]! =label[pos])
        {
            label[center] = WATERSHED;
        }
    }
}
```

/ **
涨水
参数：
　　　　double lambda 参数 λ
　　　　double beta 参数 β
　　　　vector<pair<int, int> > &seedPoints 种子点
　　　　double * gradient 影像梯度
　　　　int winsize 窗口大小
　　　　int rows 影像行数
　　　　int cols 影像列数
　　　　int * label 编号
返回值：无
* **/
```
void waterFill(double lambda, double beta, vector<pair<int, int> > &seedPoints, double * gradient, int winsize, int rows, int cols, int * label)
```

```cpp
    {
        int numPixel = rows * cols;
        //第一行和最后一行
        int i=0, j=0;
        for (i=0; i<cols; i++)
        {
            label[i] = WATERSHED;
            label[numPixel-1-i] = WATERSHED;
        }
        //第一列和最后一列
        for (i=0; i<rows; i++)
        {
            label[i * cols] = WATERSHED;
            label[(i+1) * cols-1] = WATERSHED;
        }
        //形状约束函数(模型)
        float m_bShapeConstraint[500];
        memset(m_bShapeConstraint, 0, 500 * sizeof(float));
        double shp[500];
        shp[0] = 2.8f;
        shp[1] = 2.3f;
        shp[2] = 1.9f;
        shp[3] = 1.6f;
        double maxValue = 0;
        for (i=0; i<500; i++)
        {
            m_bShapeConstraint[i] = exp(-(double)i/beta) * lambda;
        }
        unsigned char * isProcessed = new unsigned char[numPixel];
        memset(isProcessed, 0, numPixel * sizeof(unsigned char));
        std::queue<pair<int, int> > priQueue[500]; //
        //将靠近区域边缘的所有未处理点放入到相应的优先级队列
        int pos=0, rowFirstPos=0;
        pair<int, int> tempPoint;
        for (i=1; i<rows-1; i++)
        {
            rowFirstPos=i * cols;
            for (j=1; j<cols-1; j++)
```

```cpp
            {
                pos = rowFirstPos+j;
                if (label[pos]<WATERSHED)//该点未处理
                {
                    if(label[pos-1]>WATERSHED || label[pos+1]>WATERSHED ||
label[pos+cols]>WATERSHED || label[pos-cols]>WATERSHED)//四邻域有已标记点
                    {
                        tempPoint.first = i;
                        tempPoint.second = j;
                        priQueue[int(gradient[pos])].push(tempPoint);
                        isProcessed[pos] = 1;
                    }
                }
            }
        }
    }
//依次处理这256个优先级队列
unsigned short index[500];//用以记录当前梯度级的像素点该压入哪一个队列
for (i=0; i<500; i++)
{
    index[i] = i;
}
pair<int, int> seed;
pair<int, int> tempPoint2;
int pos2 = 0;
int offsetPos[4] = {-1, 1, -cols, cols};
for (i=0; i<500; i++)
{
    //将梯度级 i 之前的索引强制改变为 i
    for (j=0; j<i; j++)
    {
        index[j] = i;
    }
    double dis2seed = 0;
    while(!priQueue[i].empty())
    {
        tempPoint = priQueue[i].front();
        priQueue[i].pop();
        pos = tempPoint.first * cols+tempPoint.second;
```

```cpp
            //八邻域扩张
            for (j=0; j<4; ++j)
            {
                expandBasin(pos, pos+offsetPos[j], label);
            }
//经过扩张之后如果当前点被标记，则将其邻域点加入到对应的队列中
            if (label[pos]>=0)
            {
                seed.first = seedPoints[label[pos]].first; //获取该点的水平种子位置
                seed.second = seedPoints[label[pos]].second;
                for (j=0; j<4; j++)
                {
                    pos2=pos+offsetPos[j];
                    //当前点未赋区域标识(包括分水线)
                    if (label[pos2]<WATERSHED&&!isProcessed[pos2])//左侧点未标记
                    {
                        tempPoint2.first=pos2/cols;
                        tempPoint2.second=pos2%cols;
                        dis2seed
=(seed.first-tempPoint2.first)*(seed.first-tempPoint2.first)
+(seed.second-tempPoint2.second)*(seed.second-tempPoint2.second);
                        dis2seed = sqrt(dis2seed);
                        dis2seed = m_bShapeConstraint[i]*dis2seed;
priQueue[(int)(index[int(gradient[pos2])]+dis2seed)].push(tempPoint2);
                        isProcessed[pos2] = 1;
                    }
                }
            }
        }
//把当前所有梯度小于等于i的点全部置为i+1
    }
    delete []isProcessed;
    isProcessed = NULL;
}

/****************************************************************
影像边缘处理
参数:
    IplImage *img 输入影像
```

  int *label 编号

  vector<pair<int, int> > &seedPoints 种子点

返回值：无

* **/

```cpp
void procEdgePixs(IplImage *img, int *label, vector<pair<int, int> > &seedPoints)
{
    int rows = img->height;
    int cols = img->width;
    int channels = img->nChannels;
    int offsetPos[4] = {-1, 1, -cols, cols};  //循环变量，用于控制寻找4邻域点
    int i = 0, j = 0, k = 0, m = 0;
    int rowStart = 0, pos = 0, pos2 = 0;
    pair<int, int> tempPoint2;
    //将所有的分水线点寻找其4邻域中与之最接近的点，然后将该点的区域编号赋予它
    double dist = 0, delta = 0;
    double min = 1000000;
    int local;
    for (i = 1; i < rows-1; i++)
    {
        rowStart = i * cols;
        for (j = 1; j < cols-1; j++)
        {
            pos = rowStart+j;
            if (label[pos] < 0)
            {
                min = 1000000;
                local = 0;
                for (k = 0; k < 4; k++)
                {
                    pos2 = pos+offsetPos[k];
                    if (label[pos2] < 0)
                    {
                        dist = 0;
                        tempPoint2.first = pos2/cols;
                        tempPoint2.second = pos2%cols;
                        for (m = 0; m < channels; m++)
                        {
```

```cpp
            {delta=double(CV_IMAGE_ELEM(img, uchar, i, j*channels+m)-CV_IMAGE
_ELEM(img, uchar, tempPoint2.first, tempPoint2.second+m));
                    dist+=delta*delta;
                    }
                    delta=seedPoints[label[pos2]].first-i;
                    dist+=delta*delta;
                    delta=seedPoints[label[pos2]].second-j;
                    dist+=delta*delta;
                    dist=sqrt(dist);
                    if (dist<min)
                    {
                        local=k;
                        min=dist;
                    }
                }
            }
            label[pos]=label[pos2];
            if (label[pos]<0)
            {
                for (k=0; k<4; k++)
                {
                    if (label[pos+offsetPos[k]]>=0)
                    {
                        label[pos] = label[pos+offsetPos[k]];
                        break;
                    }
                }
            }
        }
    }
//处理四周边框
//上下边缘
//左右边缘
    for (i=0; i<rows; i++)
    {
        label[i*cols] = label[i*cols+1];
        label[(i+1)*cols-1] = label[(i+1)*cols-2];
```

```
    }
    int numPixel = rows * cols;
    for (i=0; i<cols; i++)
    {
        label[i] = label[i+cols];
        label[numPixel-1-i] = label[numPixel-1-i-cols];
    }
}
/****************************************************************
```
影像边缘再处理,如果一个像素点四邻域中仅有一个点与该点的区域号相同,则该点为毛刺,将其编号改为其他三邻域中较多那点的编号

参数:

    int *label 编号

    int rows 影像行数

    int cols 影像列数

返回值:无

```
*****************************************************************/
void smootBoundaries(int *label, int rows, int cols)
{
    int up=0, down=0, left=0, right=0;
    int flag=0;
    int m_b[4]={0};
    int i=0, j=0, rowStart=0, pos=0;
    for (i=1; i<rows-1; i++)
    {
        rowStart=i*cols;
        for (j=1; j<cols-1; j++)
        {
            pos=rowStart+j;
            up=pos-cols;
            down=pos+cols;
            left=pos-1;
            right=pos+1;
            flag = 0;
            memset(m_b, 0, 4*sizeof(int));
            if (label[pos]==label[up])   { flag++; m_b[0] = 1;}
            if (label[pos]==label[left]) { flag++; m_b[1] = 1;}
            if (label[pos]==label[down]) { flag++; m_b[2] = 1;}
```

```cpp
                if(label[pos]==label[right]) {flag++; m_b[3] = 1;}
                if(flag==1)//四邻域仅有一个点与之有相同标记
                {
                    if(label[up]==label[down] || label[left]==label[right])
                    {
                        if(!m_b[0]) {label[pos] = label[up];}
                        else if(!m_b[1]) {label[pos] = label[left];}
                        else if(!m_b[2]) {label[pos] = label[down];}
                        else if(!m_b[3]) {label[pos] = label[right];}
                    }
                }
            }
        }
    }
}
/**************************************************************************
设置编号
参数：
    IplImage *img 输入影像
    int *label 编号
返回值：无
***************************************************************************/
void setColor(IplImage *img, int *label)
{
    int i=0, j=0, k=0;
    int rows=img->height;
    int cols=img->width;
    int channles=img->nChannels;
    for(i = 0; i <rows; i++)
    {
        for(j = 0; j <cols; j++)
        {
            if(label[i*cols+j]<0)
            {
                for(k=0; k<channles-1; k++)
                {
                    CV_IMAGE_ELEM(img, uchar, i, j*channles+k)= 0;
                }
                CV_IMAGE_ELEM(img, uchar, i, j*channles+k)= 255;
```

```cpp
                }
            }
        }
}
/************************************************************
Sections 类
*************************************************************/
class Sections
{
public:
    void findSections(int *label, int rows, int cols);
    void showSection(IplImage *img, int row, int col);

private:
    std::vector<std::vector<std::pair<int, int> > > sections;
    std::vector<int> data;
    std::vector<std::vector<int> > ave;
};
/************************************************************
寻找区域
参数:
    int *label 编号
    int rows 影像行数
    int cols 影像列数
返回值: 无
*************************************************************/
void Sections::findSections(int *label, int rows, int cols)
{
    sections.clear();
    data.clear();
    int i=0, j=0, k=0;
    int maxSecID=0;
    int numLabel=rows*cols;
    data.resize(numLabel);
    for (i=0; i<numLabel; ++i)
    {
        data[i]=label[i];
        if (label[i]>maxSecID)
```

```
                }
                    maxSecID = label[i];
            }
        }
        int numSec = maxSecID+1;
        sections.resize(numSec);
        int offsetPos[4] = {-1, 1, -cols, cols};
        int pos = 0, rowStart = 0, pos2 = 0;
        pair<int, int> tempPoint;
        for (i=1; i<rows-1; ++i)
        {
            rowStart = i * cols;
            for (j=1; j<cols-1; ++j)
            {
                pos = rowStart+j;
                if (label[pos]<0)
                {
                    for (k=0; k<4; ++k)
                    {
                        pos2 = pos+offsetPos[k];
                        if (label[pos2]>=0)
                        {
                            tempPoint.first = i;
                            tempPoint.second = j;
                            sections[label[pos2]].push_back(tempPoint);
                        }
                    }
                }
            }
        }
    }
```

/***

显示区域

参数：

    IplImage *img 输入影像

    int rows 影像行数

    int cols 影像列数

返回值：无

```cpp
* *********************************************************/
void Sections::showSection(IplImage *img, int row, int col)
{
    int pos=row*img->width+col;
    int secID=data[pos];
    if (secID<0)
    {
        int offsetPos[4]={-1, 1, -img->width, img->width};
        int numPixel=img->width * img->height;
        int pos2=0;
        for (int i=0; i<4; ++i)
        {
            pos2=pos+offsetPos[i];
            if (pos2>=0 && pos2<numPixel && data[pos2]>0)
            {
                pos=pos2;
                secID=data[pos];
                continue;
            }
        }
        if (secID<0)
        {
            return;
        }
    }
    int i=0, j=0, k=0;
    int channels=img->nChannels;
    vector<pair<int, int> >::iterator ite;
    for (ite=sections[secID].begin(); ite!=sections[secID].end(); ++ite)
    {
        for (i=0; i<channels-1; ++i)
        {
            CV_IMAGE_ELEM(img, uchar, ite->first, ite->second * channels+i)=0;
        }
        CV_IMAGE_ELEM(img, uchar, ite->first, ite->second * channels+i)=255;
    }
}
```

参 考 文 献

1. 贾永红. 数字图像处理(第二版). 武汉：武汉大学出版社，2010.
2. 求是科技. Visual C++数字图像处理典型算法及实现. 北京：人民邮电出版社，2006.
3. 何斌，马天予，王运坚，朱红莲. VC++数字图像处理. 北京：人民邮电出版社，2001.
4. 杨淑莹. 图像模式识别 VC++技术实现，北京：清华大学出版社，北京交通大学出版社，2005.
5. 李民录. GDAL 源码剖析与开发指南. 北京：人民邮电出版社，2014.
6. 于仕琪，刘瑞祯. 学习 OpenCV(中文版). 北京：清华大学出版社. 2009.
7. 参阅网站：

    http：//www. gdal. org/index. html

    http：//www. opencv. org/

    http：//www. openmp. org/

    http：//www. openrs. org/